Arduino

LCD

Projects

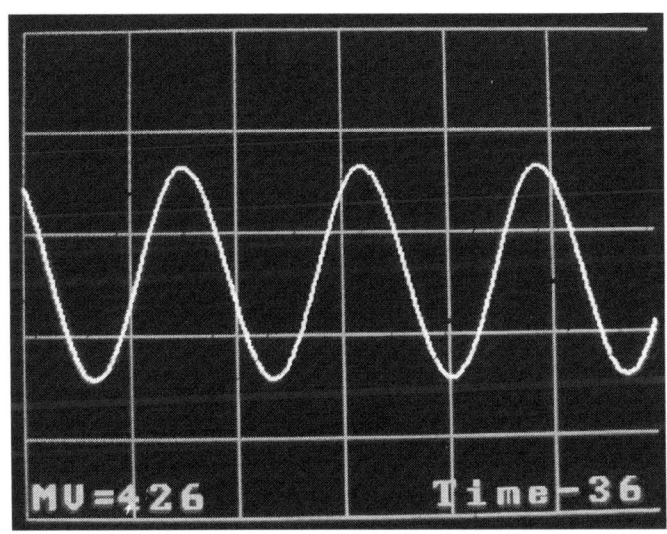

Robert J Davis II

Arduino LCD Projects

Copyright 2013 by Robert J Davis
All Rights Reserved.
Copying by permission only, except for short quotes.

This book is the second in my series of Arduino projects books. The first one was "Arduino LED Projects". My plan is to write several additional books in this series. Each book will be more complex than the preceding book, as the books progress.

As usual, the construction and safe operation of all of these devices is totally the responsibility of the reader or builder. If you do not know what you are doing, do not try these projects. You might want to start with the simpler projects that are found in my "Arduino LED Projects" book.

Every effort is made to make these projects as simple and as easy to build as is possible. Fun and practical applications are given for each project. Feel free to add your own customizations and modifications to make even better projects.

Each project in this book has a quick explanation, a picture of the working project, a schematic, and a software or sketch listing. This book is only meant to get you started at building some Arduino based LCD projects. Feel free to modify, improve, or even "play" with the software and hardware. Electronics can be lots of fun, and that is what you should do with these projects. Have some fun!

This book is intended as a reference guide to some of the most common LCD displays that can be used with an Arduino processor. As such, I will give examples of both text and graphics based LCD's and how to set them up and use them with some sample applications. There is no way that I could possibly include all of the many available LCD's. The LCD's that are shown here should give you enough information to be able to use some of the other LCD's as well.

Table of Contents:

1. Introduction to LCD's……………..………………………. 4
2. Installing the Arduino Drivers…………..………………… 6
3. Single Line LCD 16 by 1, Model 16108……………………. 8
    Temperature Display
4. Two Line LCD 16 by 2, Model QC1602A……………….. 12
    Indoor/Outdoor temperature display
    Simulated analog meter
    Bar graph meter
5. Two Line LCD 40 by 2, Model DMC50037N………………. 21
    Six analog inputs
    Stereo bar graph meter
6. Four Line LCD 20 by 4, Model J204A……………………….. 28
    Four temperature display
    Dual simulated analog meter
7. Low Resolution Graphics LCD, 48x84, Nokia 5110…………. 34
    Text display test program
    Six analog inputs displayed
    Simple Arduino oscilloscope
    Analog meter simulation
8. Medium Resolution Graphics LCD, 128x64, QC12864B…... 46
    Analog meter simulation
    Better Arduino oscilloscope
    Two channel logic analyzer
    Six channel logic analyzer
    External analog to digital converter oscilloscope
9. Medium Resolution Graphics LCD, 160x128………………... 71
    (1.8TFT SPI)
    Analog meter simulation
    Color Arduino oscilloscope
10. High Resolution Graphics LCD, 320x240………………….. 78
    (TFT240_262K, 2.4")
    Three color analog meter
    Three color oscilloscope
    Two channel logic analyzer
    Six channel logic analyzer
    External analog to digital converter oscilloscope
Bibliography…………………………………………………….104

Chapter 1

Introduction to LCD's

LCD stands for "Liquid Crystal Display". These displays contain a liquid that changes states from a liquid to a crystal (Solid) when a small voltage is applied. In the liquid state it is transparent and in the crystal state it reflects light. An optional LED backlight then makes the LCD readable in dim light or where there is no light to reflect.

The Arduino programmers wrote a library of drivers for LCD "text" type of displays. The library adds several new commands that you can now use to communicate with the LCD display. You will need to include that library in your LCD programs. This is the code to include the Liquid Crystal library.

```
// include the library code:
#include <LiquidCrystal.h>
```

Here are some of the LCD Liquid Crystal library commands for the text type of LCD's:

lcd.begin(C,R); where C = Column, and R = Row.
Sets up, and specifies the width and height of the LCD display.

lcd.setCursor(C,R); where C = Column, and R = Row.
Basically you are telling the display where you want to write text to.

lcd.print("test"); where "test" is some text to be displayed by the LCD. This is how you send things to the display so it can be read by you.

lcd.clear(); Clears out whatever is in the display.

lcd.home(); Moves the cursor to 0,0 or the "home" position.

lcd.cursor(); Displays an underline "cursor" at the current position.

lcd.noCursor(); Turns off the underline "cursor".

lcd.blink(); Makes the cursor blink.

lcd.noBlink(); Turns off the blinking of the cursor..

Chapter 2

Installing the Arduino Drivers

In my last book I used Arduino drivers that are now out of date. To install the drivers first decompress the files into a known location. I like to use the root directory of the C: drive.

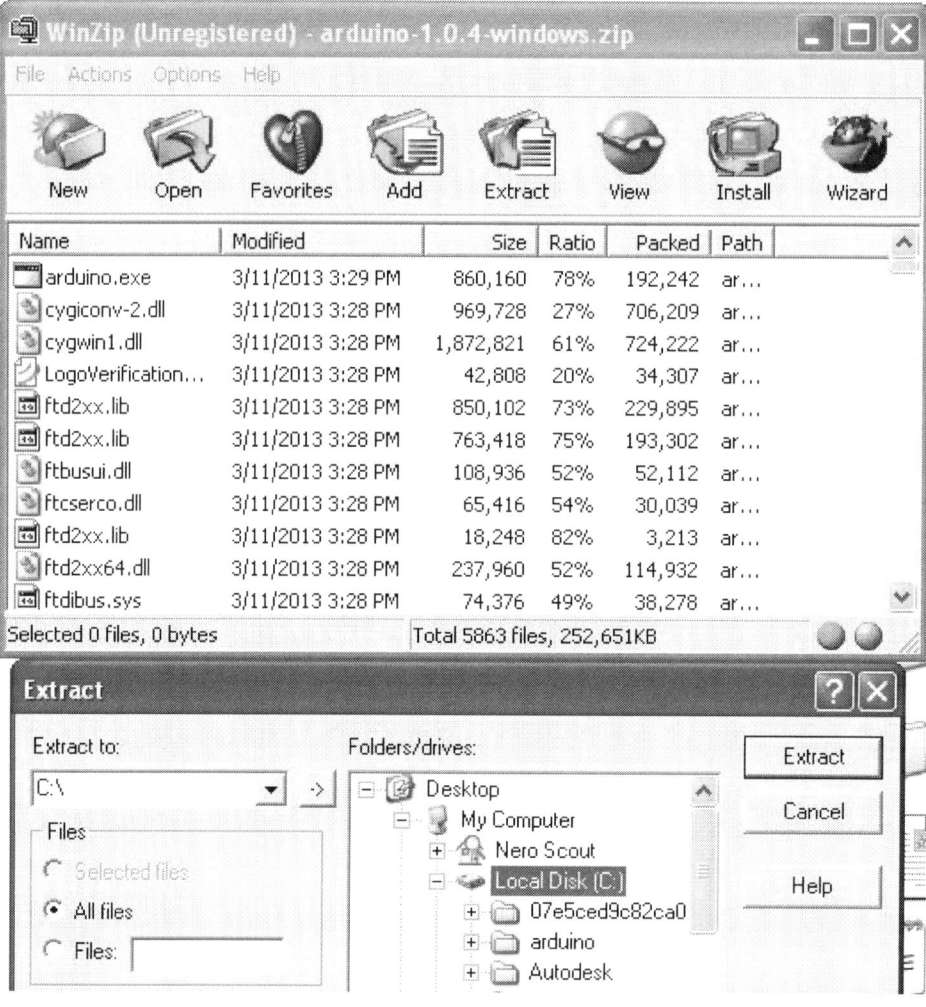

The decompressed files create a directory called "Arduino-1.0.4" with all of the needed files. You can now plug in the Arduino UNO with a USB cable. Windows should come up with the "Install new hardware" wizard. You will need to tell it to "Install from a specific location", and then direct it to the "Arduino-1.0.4\drivers" directory. Everything should now install and Windows will then recognize the Arduino.

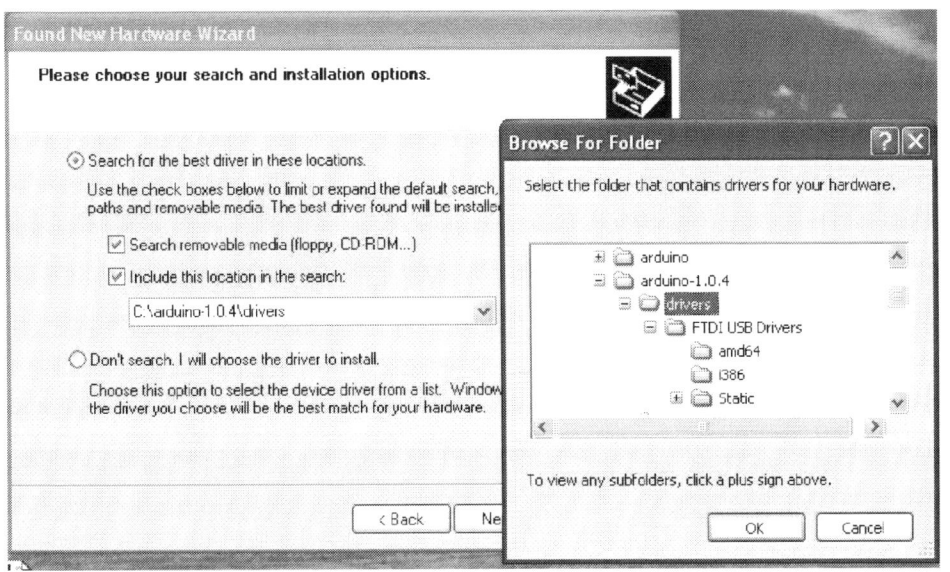

Also in the Arduino-1.0.4 directory there is a program called "Arduino.exe". It is the program that you will use to communicate with and upload your sketches to the Arduino. It is a good idea to create a shortcut to it on your computers desktop so that it is easy to find.

There is also a subdirectory of Arduino-1.0.4 that is called "reference". In that subdirectory there is a file called "index.html". When you double click on that file a reference library should come up in your favorite browser. This library can be very helpful in debugging your code. It can also be accessed within the Arduino.exe program by selecting "help" and "reference".

Chapter 3

Single Line LCD 16108 – 16 by 1

Our first project will be to make a single line LCD display come to life. To do that, we will need two control lines, four data lines, 5V power and ground. We will also need a variable resistor to adjust the voltage on the "contrast" pin. I usually end up adjusting the contrast control almost all the way to one end, to get the LCD to work.

Here are the pin assignments for a typical text based LCD display. This pinout stays the same for most text displays that use a Hitachi HD44780 or similar LCD display controller chip.

Pin	Name	Function
1	Vss	Ground
2	Vdd	Power + 5 volts
3	Vo	Contrast Adjustment
4	RS	Register Select
5	R/W	Data Read Write
6	E	Enable Signal
7	DB0	Data Buss Bit 0
8	DB1	Data Buss Bit 1
9	DB2	Data Buss Bit 2
10	DB3	Data Buss Bit 3
11	DB4	Data Buss Bit 4
12	DB5	Data Buss Bit 5
13	DB6	Data Buss Bit 6
14	DB7	Data Buss Bit 7
15	A	LCD Backlight Positive
16	K	LCD Backlight Ground

The next picture is a picture of what the LCD looks like once it is properly wired up and running. To get the wiring to look this simple, you will need to add jumper wires that cross over the center barrier of the breadboard.

Otherwise the LCD is right on top of the wiring. The jumpers are hidden underneath the LCD display in the picture.

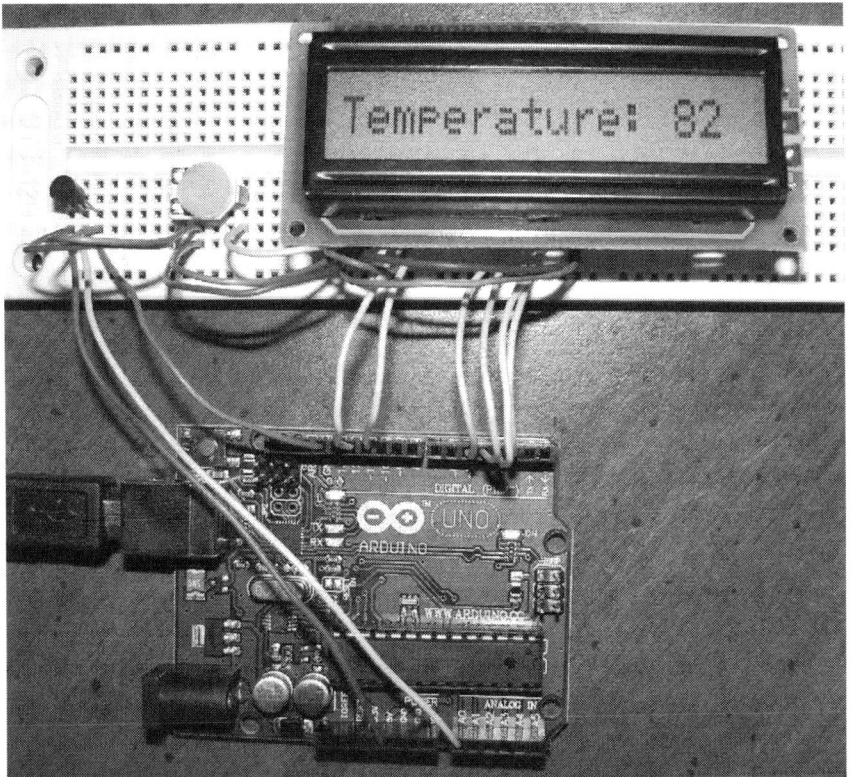

Here is the schematic diagram of how to wire the LCD up to the Arduino UNO. The LCD display in use has the pins at the top of the display. I used jumpers not shown in the schematic that jump over the center of the breadboard to make it easier to wire up the LCD display.

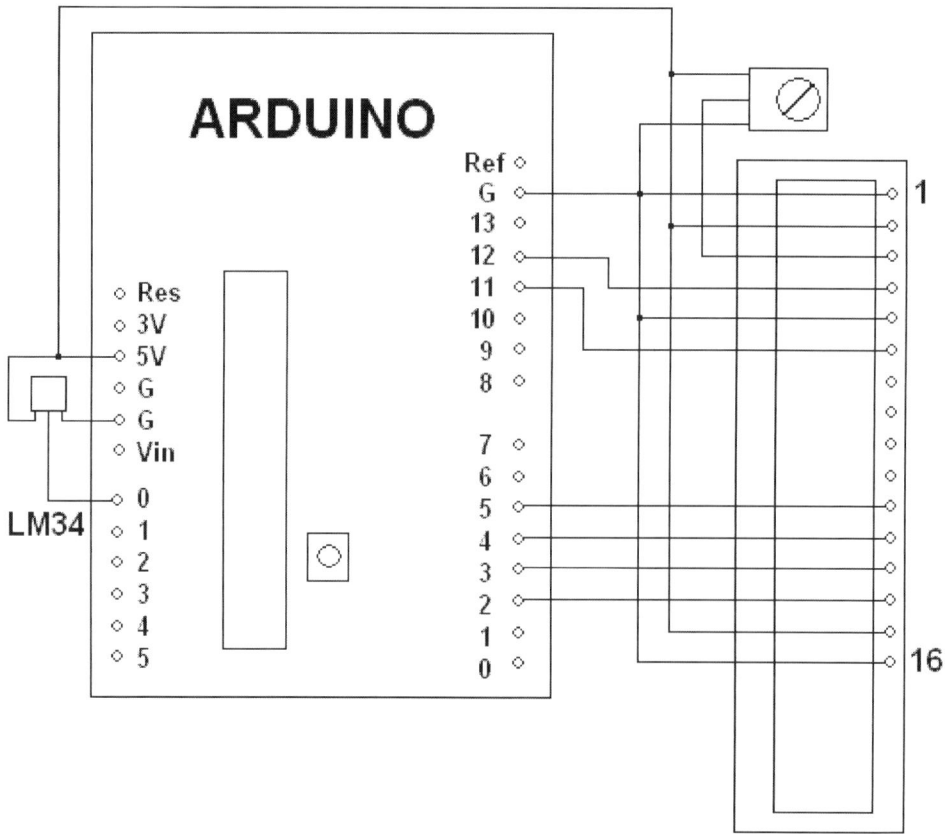

Here is the sketch or code to make it display the temperature from a LM34 connected to Analog 0.

/***
 Liquid Crystal Library - Temperature
 Demonstrates the use a 16x1 LCD display.
 It is organized as row 0 has 8 characters,
 and row 1 has 8 characters
 This sketch prints "Temperature" to the LCD
 and shows the temperature.

 The circuit:
 * LCD RS pin to digital pin 12
 * LCD En pin to digital pin 11
 * LCD D4 pin to digital pin 5
 * LCD D5 pin to digital pin 4
 * LCD D6 pin to digital pin 3
 * LCD D7 pin to digital pin 2

* LCD R/W pin to ground
* Variable resistor wiper to LCD VO pin (pin 3)
Adapted from code written by David A. Mellis
Modified by Limor Fried, Tom Igoe, Bob Davis
***/

```
// include the library code:
#include <LiquidCrystal.h>
// initialize the library with interface pin #'s
LiquidCrystal lcd(12, 11, 5, 4, 3, 2);

void setup() {
  // set up the LCD's number of columns and rows:
  lcd.begin(8, 2);
  // Print the first 8 characters to the LCD.
  lcd.setCursor(0, 0);
  lcd.print("Temperat");
  // Print the last 3 characters to the LCD.
  lcd.setCursor(0, 1);
  lcd.print("ure:");
}

void loop() {
  // set the cursor to column 5 of line 1
  lcd.setCursor(5, 1);
  int Temperature=analogRead(A0)/2.05;
  // print the temperature:
  lcd.print(Temperature);
}
```

Chapter 4

Two Line LCD 1602A - 16 by 2

Our next project will be to make a two line LCD display operational. To do that, we will need everything that we used for the single line LCD. You can literally unplug the one line LCD, plug in the two line LCD, and everything should work. Most of the changes are in the code or sketch. We will also add an additional input of a second temperature sensor, for one project and a variable resistor, for another project.

Here is a picture of the indoor/outdoor temperature display while it is working. The difference between the temperatures is from me squeezing the outdoor temperature sensor, thus running up that temperature.

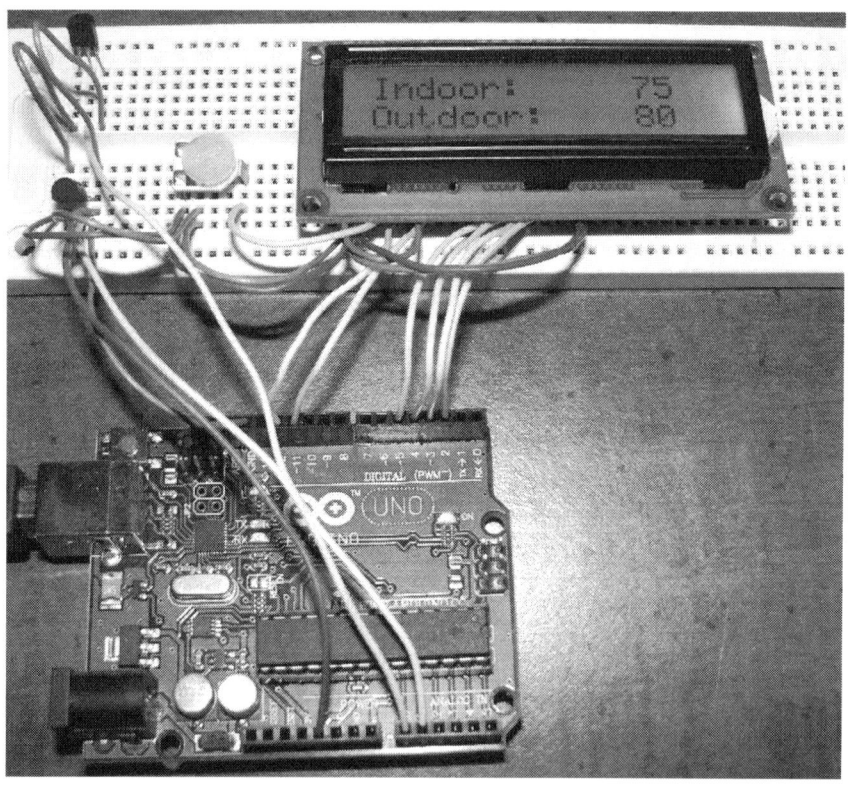

Below is the schematic diagram of the two line LCD interface. It is mostly the same as the one line LCD except for the second, additional temperature sensor that has been added.

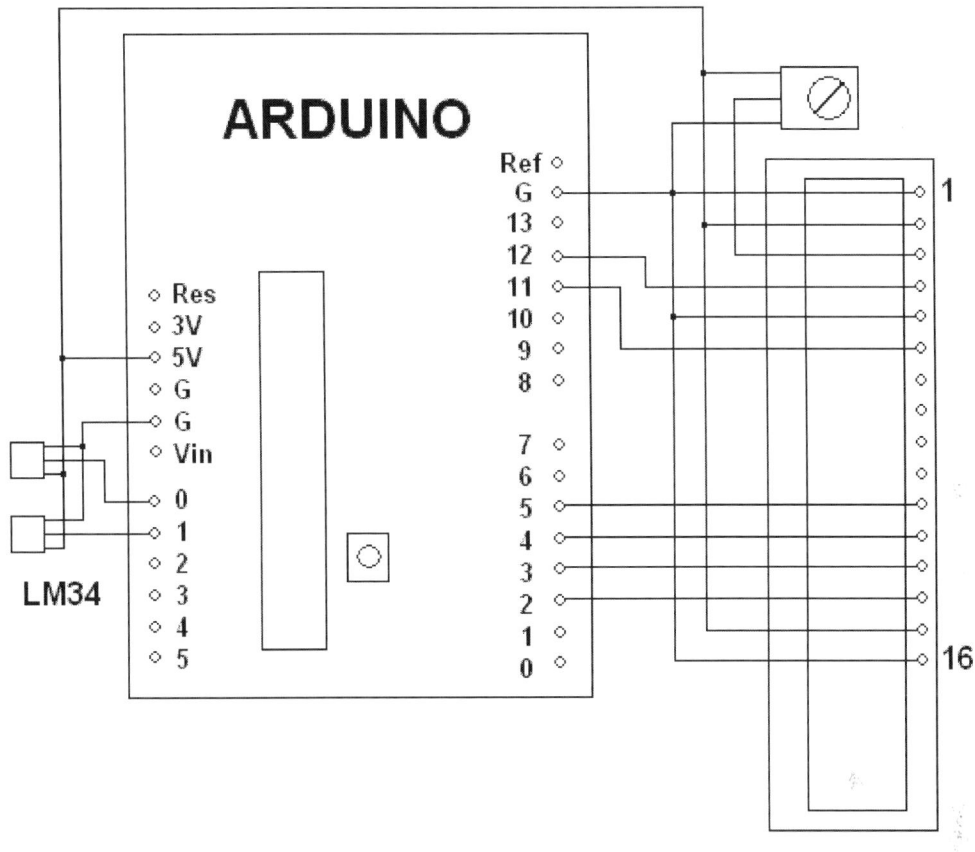

Here is the sketch or code for an indoor/outdoor thermometer display.

/***
 Liquid Crystal Library - Temperature
 Demonstrates the use a 16x2 LCD display.
 This sketch prints "Indoor" and "outdoor" to the LCD
 and then shows the temperatures.

 The circuit:
 * LCD RS pin to digital pin 12
 * LCD En pin to digital pin 11
 * LCD D4 pin to digital pin 5
 * LCD D5 pin to digital pin 4
 * LCD D6 pin to digital pin 3

* LCD D7 pin to digital pin 2
* LCD R/W pin to ground
* Variable resistor wiper to LCD VO pin (pin 3)

Adapted from code written by David A. Mellis
Modified by Limor Fried, Tom Igoe, Bob Davis
**/

```
// include the library code:
#include <LiquidCrystal.h>

// initialize the library with interface pin #'s
LiquidCrystal lcd(12, 11, 5, 4, 3, 2);

void setup() {
  // set up the LCD's number of columns and rows:
  lcd.begin(16, 2);
  // Print the text to the LCD.
  lcd.setCursor(0, 0);
  lcd.print("Indoor:");
  lcd.setCursor(0, 1);
  lcd.print("Outdoor:");
}

void loop() {
  // set the cursor to column 12 of line 0
  lcd.setCursor(12, 0);
  int Temperature=analogRead(A0)/2.05;
  // print the temperature:
  lcd.print(Temperature);
  // set the cursor to column 12 of line 1
  lcd.setCursor(12, 1);
  int Temperature2=analogRead(A1)/2.05;
  // print the temperature:
  lcd.print(Temperature2);
}
```

The next picture is of the two line LCD, with a simulated analog meter display. The meter needle was created by using custom characters. The custom characters are called line1, line2, line3, line4, and line5. They are vertical lines that are designed to look like the needle of an analog meter.

In the picture, you cannot see the line moving as the voltage is turned up or down. Due to the math involved, the maximum voltage that can displayed, on the simulated analog meter, is 4.8 volts.

The meter is limited to 4.8 volts because first we divided the input, which has a maximum of 1024, by 2.05 to get a maximum of 500 millivolts. Then we divide that result by three to get a maximum of 166.6. Next we divide that number by 10 to get a maximum of 16.66. There are 16 available character positions, so the last .66 is lost as off the top of our meters scale. Thus the meter maxes out at 4.8 volts.

Here is the code for what looks like an analog meter. It uses custom characters to create five lines for the five positions, inside of each character position. I used a variable resistor for the input to test this sketch. That gave an input voltage that varied from zero volts to five volts.

/***
 Liquid Crystal Library - Analog meter
 Demonstrates the use a 16x2 LCD display.
 This sketch displays a simulated analog meter to the LCD

 The circuit:
 * LCD RS pin to digital pin 12
 * LCD En pin to digital pin 11
 * LCD D4 pin to digital pin 5
 * LCD D5 pin to digital pin 4
 * LCD D6 pin to digital pin 3
 * LCD D7 pin to digital pin 2
 * LCD R/W pin to ground
 * Variable resistor wiper to LCD VO pin (pin 3)

 Adapted from code written by David A. Mellis
 Modified by Limor Fried, Tom Igoe, Bob Davis
 ***/
// include the library code:
#include <LiquidCrystal.h>
// initialize the library with interface pin #'s
LiquidCrystal lcd(12, 11, 5, 4, 3, 2);

byte line1[8]= {
  B10000, B10000, B10000, B10000,
  B10000, B10000, B10000, B10000
};
byte line2[8]= {
  B01000, B01000, B01000, B01000,
  B01000, B01000, B01000, B01000
};
byte line3[8]= {
  B00100, B00100, B00100, B00100,
  B00100, B00100, B00100, B00100
};
byte line4[8]= {
  B00010, B00010, B00010, B00010,

```
  B00010, B00010, B00010, B00010
};
byte line5[8]= {
  B00001, B00001, B00001, B00001,
  B00001, B00001, B00001, B00001
};

void setup() {
  // set up the LCD's number of columns and rows:
  lcd.begin(16, 2);
  // Print the text to the LCD.
  lcd.createChar(1, line1);
  lcd.createChar(2, line2);
  lcd.createChar(3, line3);
  lcd.createChar(4, line4);
  lcd.createChar(5, line5);
}

void loop() {
  lcd.clear();
  lcd.setCursor(0, 0);
  lcd.print("MiliVolts:");
  // set the cursor to column 12 of line 0
  lcd.setCursor(12, 0);
  int Voltage=analogRead(A0)/2.05;
  // print the voltage:
  lcd.print(Voltage);
  int metervolts=Voltage/3;
  lcd.setCursor(metervolts/10, 1);
  if (metervolts%10/2 == 0) lcd.write(1);
  if (metervolts%10/2 == 1) lcd.write(2);
  if (metervolts%10/2 == 2) lcd.write(3);
  if (metervolts%10/2 == 3) lcd.write(4);
  if (metervolts%10/2 == 4) lcd.write(5);
  delay(100);
}
```

You can take the simulated analog meter a step further, and make it into a bar graph type of meter. It is just a matter of changing the "lines" into "blocks", and then drawing the blocks below the current position as solids. Here is a picture of the bar graph LCD display.

Here is the sketch or code to make the bar graph meter display work.

/***

LiquidCrystal Library - Bar Graph
Demonstrates the use a 16x2 LCD display.
This sketch displays a simulated bar graph to the LCD

 The circuit:
* LCD RS pin to digital pin 12
* LCD En pin to digital pin 11
* LCD D4 pin to digital pin 5
* LCD D5 pin to digital pin 4
* LCD D6 pin to digital pin 3
* LCD D7 pin to digital pin 2
* LCD R/W pin to ground
* Variable resistor wiper to LCD VO pin (pin 3)

Adapted from code written by David A. Mellis
Modified by Limor Fried, Tom Igoe, Bob Davis
***/

```
// include the library code:
#include <LiquidCrystal.h>
// initialize the library with interface pin #'s
LiquidCrystal lcd(12, 11, 5, 4, 3, 2);

byte line1[8]= {
  B10000, B10000, B10000, B10000,
  B10000, B10000, B10000, B10000
};
byte line2[8]= {
  B11000, B11000, B11000, B11000,
  B11000, B11000, B11000, B11000
};
byte line3[8]= {
  B11100, B11100, B11100, B11100,
  B11100, B11100, B11100, B11100
};
byte line4[8]= {
  B11110, B11110, B11110, B11110,
  B11110, B11110, B11110, B11110
};
byte line5[8]= {
  B11111, B11111, B11111, B11111,
  B11111, B11111, B11111, B11111
};

void setup() {
  // set up the LCD's number of columns and rows:
  lcd.begin(16, 2);
  // create the text for the LCD.
  lcd.createChar(1, line1);
  lcd.createChar(2, line2);
  lcd.createChar(3, line3);
  lcd.createChar(4, line4);
  lcd.createChar(5, line5);
}

void loop() {
```

```
lcd.clear();
lcd.setCursor(0,0);
lcd.print("Volts:");
int Voltage=analogRead(A0)/2.05;
// print the voltage:
lcd.setCursor(8, 0);
lcd.print(Voltage);
// overwrite first digit and add a period
lcd.setCursor(7, 0);
if (Voltage >= 100) lcd.print(Voltage/100);
lcd.print(".");
if (Voltage < 100) lcd.print(Voltage);
// print the bar graph
lcd.setCursor(0,1);
int metervolts=Voltage/3;
// draw lower spaces as solid
for (int pos=0; pos < metervolts/10; pos++) {
lcd.write(5);
}
lcd.setCursor(metervolts/10, 1);
if (metervolts%10/2 == 0) lcd.write(1);
if (metervolts%10/2 == 1) lcd.write(2);
if (metervolts%10/2 == 2) lcd.write(3);
if (metervolts%10/2 == 3) lcd.write(4);
if (metervolts%10/2 == 4) lcd.write(5);
delay(500);
}
```

Chapter 5

Two Line LCD DMC50037N - 40 by 2

Our next project will be to make a two line by 40 character LCD display work. To do that, we will need everything that we used for the previous two line LCD, and some ribbon cable jumper wires. You can usually easily unplug one text based LCD, and plug in another text based LCD. However, this time, I chose this LCD because it does not have the usual 16 pin in line connector. Instead, this LCD has a 14 pin connector configured in two rows of 7 pins each. Pins 15 and 16 are not there because it has no backlight LED.

The schematic diagram is the same as the one we used for the previous two line LCD display. The only change is that there is no connection to pins 15 and 16 since they do not exist.

Here is a close up picture of the connector, with a ribbon cable connected to it. I skipped over the four unused pins that are numbered seven to ten.

Below is a picture of the LCD displaying the contents of all six analog inputs. Note that the jumpers over the center of the breadboard are now visible. Also, this LCD does not have a backlight so there are no connections to pins 15 or 16.

Here is the sketch for seeing the status of all of the six analog inputs.

/***
 LiquidCrystal Library - 40 x 2 - 6 analog
 Demonstrates the use a 40x2 LCD display.
 This sketch prints "Analog0" to "Analog5" to the LCD
 and then it shows the analog readings.

 The circuit:
* LCD RS pin to digital pin 12
* LCD En pin to digital pin 11
* LCD D4 pin to digital pin 5
* LCD D5 pin to digital pin 4
* LCD D6 pin to digital pin 3
* LCD D7 pin to digital pin 2
* LCD R/W pin to ground

* Variable resistor wiper to LCD VO pin (pin 3)

Adapted from code written by David A. Mellis
Modified by Limor Fried, Tom Igoe, Bob Davis
***/

```
// include the library code:
#include <LiquidCrystal.h>
// initialize the library with interface pin #'s
LiquidCrystal lcd(12, 11, 5, 4, 3, 2);

void setup() {
  // set up the LCD's number of columns and rows:
  lcd.begin(16, 2);
  // Print the text to the LCD.
  lcd.setCursor(0, 0);
  lcd.print("Analog0:   Analog2:   Analog4:");
  lcd.setCursor(0, 1);
  lcd.print("Analog1:   Analog3:   Analog5:");
}

void loop() {
  // set the cursor to column 9 of line 0
  lcd.setCursor(9, 0);
  int Analog0=analogRead(A0)/2.05;
  // print the temperature:
  lcd.print(Analog0);
  // set the cursor to column 9 of line 1
  lcd.setCursor(9, 1);
  int Analog1=analogRead(A1)/2.05;
  // print the temperature:
  lcd.print(Analog1);
  // set the cursor to column 22 of line 0
  lcd.setCursor(22, 0);
  int Analog2=analogRead(A2)/2.05;
  // print the temperature:
  lcd.print(Analog2);
  // set the cursor to column 22 of line 1
  lcd.setCursor(22, 1);
  int Analog3=analogRead(A3)/2.05;
  // print the temperature:
```

```
lcd.print(Analog3);
// set the cursor to column 35 of line 0
lcd.setCursor(35, 0);
int Analog4=analogRead(A4)/2.05;
// print the temperature:
lcd.print(Analog4);
// set the cursor to column 35 of line 1
lcd.setCursor(35, 1);
int Analog5=analogRead(A5)/2.05;
// print the temperature:
lcd.print(Analog5);
}
```

For our next sketch we will create a dual or stereo bar graph using the 40 by 2 LCD. Here is a picture of that sketch in operation.

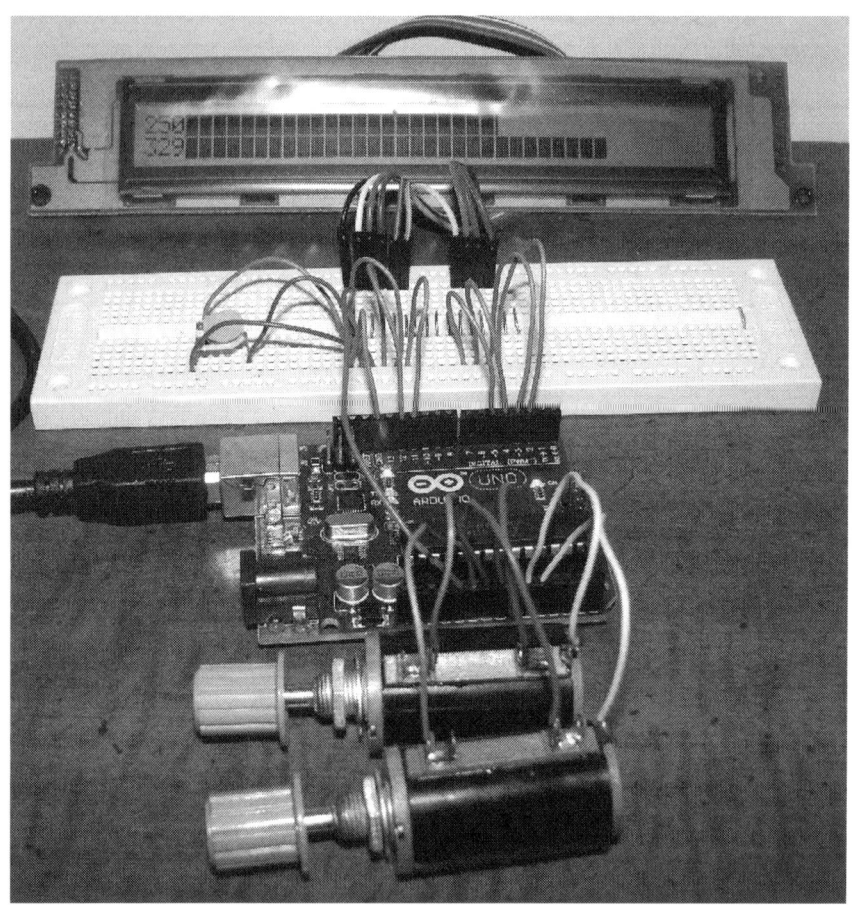

Here is the code for a sketch that will show a dual bar graph display.

```
/*****************************************
LiquidCrystal Library – Dual Bar Graph
Demonstrates the use a 40x2 LCD display.
This sketch displays a simulated bar graph to the LCD

 The circuit:
* LCD RS pin to digital pin 12
* LCD En pin to digital pin 11
* LCD D4 pin to digital pin 5
* LCD D5 pin to digital pin 4
* LCD D6 pin to digital pin 3
* LCD D7 pin to digital pin 2
* LCD R/W pin to ground
* Variable resistor wiper to LCD VO pin (pin 3)

 Adapted from code written by David A. Mellis
 Modified by Limor Fried, Tom Igoe, Bob Davis
*****************************************/
// include the library code:
#include <LiquidCrystal.h>
// initialize the library with interface pin #'s
LiquidCrystal lcd(12, 11, 5, 4, 3, 2);

byte line1[8]= {
  B10000, B10000, B10000, B10000,
  B10000, B10000, B10000, B10000
};
byte line2[8]= {
  B11000, B11000, B11000, B11000,
  B11000, B11000, B11000, B11000
};
byte line3[8]= {
  B11100, B11100, B11100, B11100,
  B11100, B11100, B11100, B11100
};
byte line4[8]= {
  B11110, B11110, B11110, B11110,
  B11110, B11110, B11110, B11110
```

```
};
byte line5[8]= {
  B11111, B11111, B11111, B11111,
  B11111, B11111, B11111, B11111
};

void setup() {
  // set up the LCD's number of columns and rows:
  lcd.begin(16, 2);
  // create the text for the LCD.
  lcd.createChar(1, line1);
  lcd.createChar(2, line2);
  lcd.createChar(3, line3);
  lcd.createChar(4, line4);
  lcd.createChar(5, line5);
}

void loop() {
  lcd.clear();
  int Voltage0=analogRead(A0)/2.05;
  int Voltage1=analogRead(A1)/2.05;
  // print the top bar graph
  lcd.setCursor(0,0);
  // draw lower spaces as solids
  for (int pos=0; pos < Voltage0/10; pos++) {
  lcd.write(5);
  }
  // print the lower bar graph
  lcd.setCursor(0,1);
  // draw lower spaces as solids
  for (int pos=0; pos < Voltage1/10; pos++) {
  lcd.write(5);
  }
  lcd.setCursor(Voltage0/10, 0);
  if (Voltage0%10/2 == 0) lcd.write(1);
  if (Voltage0%10/2 == 1) lcd.write(2);
  if (Voltage0%10/2 == 2) lcd.write(3);
  if (Voltage0%10/2 == 3) lcd.write(4);
  if (Voltage0%10/2 == 4) lcd.write(5);
  lcd.setCursor(Voltage1/10, 1);
  if (Voltage1%10/2 == 0) lcd.write(1);
```

```
  if (Voltage1%10/2 == 1) lcd.write(2);
  if (Voltage1%10/2 == 2) lcd.write(3);
  if (Voltage1%10/2 == 3) lcd.write(4);
  if (Voltage1%10/2 == 4) lcd.write(5);
  // print the voltages:
  lcd.setCursor(0, 0);
  lcd.print(Voltage0);
  lcd.setCursor(0, 1);
  lcd.print(Voltage1);
  delay(500);
}
```

Chapter 6

Four Line LCD J204A – 20 by 4

For our next project will set up a four line LCD display. These LCD's work the same way as the one and two line LCD's. The changes, once again, are mostly in the software sketch. The physical size of the LCD is also much larger than the one and the two line LCD's.

Here is a picture of the four line LCD displaying four temperatures. For the picture the indoor and outdoor inputs were tied to the same temperature sensor, and the "Hot Water" sensor was not connected, so it is reading garbage.

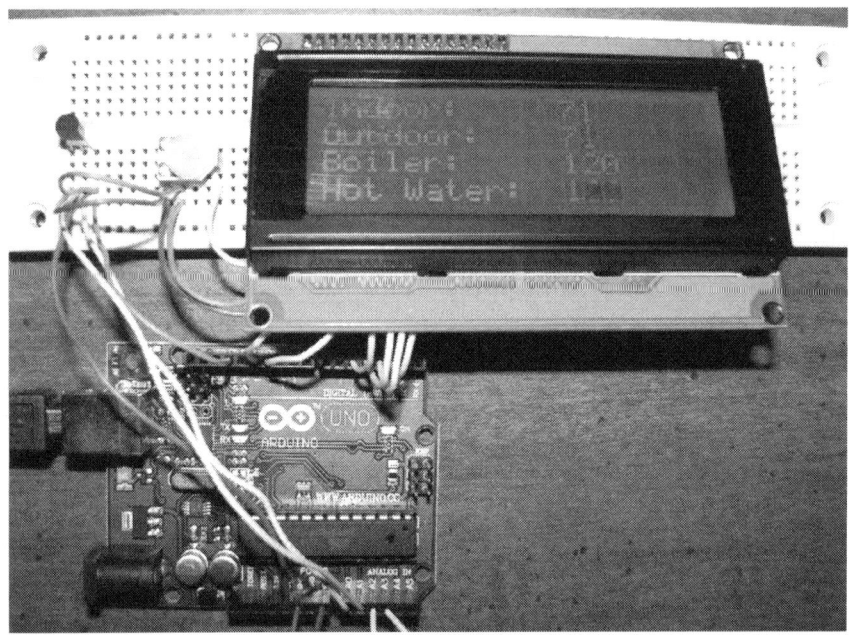

Here is the sketch for displaying four temperatures.

/***************************************
 Liquid Crystal Library - Temperature

Demonstrates the use a 20x4 LCD display.
This sketch prints text the LCD
and then it shows the temperatures.

The circuit:
* LCD RS pin to digital pin 12
* LCD En pin to digital pin 11
* LCD D4 pin to digital pin 5
* LCD D5 pin to digital pin 4
* LCD D6 pin to digital pin 3
* LCD D7 pin to digital pin 2
* LCD R/W pin to ground
* Variable resistor wiper to LCD VO pin (pin 3)

Adapted from code written by David A. Mellis
Modified by Limor Fried, Tom Igoe, Bob Davis
***/

```
// include the library code:
#include <LiquidCrystal.h>
// initialize the library with interface pin #'s
LiquidCrystal lcd(12, 11, 5, 4, 3, 2);

void setup() {
  // set up the LCD's number of columns and rows:
  lcd.begin(16, 4);
  // Print the text to the LCD.
  lcd.setCursor(0, 0);
  lcd.print("Indoor:");
  lcd.setCursor(0, 1);
  lcd.print("Outdoor:");
  lcd.setCursor(0, 2);
  lcd.print("Boiler:");
  lcd.setCursor(0, 3);
  lcd.print("Hot Water:");
}

void loop() {
  // set the cursor to column 12 of line 0
  lcd.setCursor(12, 0);
  int Temperature=analogRead(A0)/2.05;
  // print the temperature:
```

```
  lcd.print(Temperature);
  // set the cursor to column 12 of line 1
  lcd.setCursor(12, 1);
  int Temperature1=analogRead(A1)/2.05;
  // print the temperature:
  lcd.print(Temperature1);
  // set the cursor to column 12 of line 2
  lcd.setCursor(12, 2);
  int Temperature2=analogRead(A2)/2.05;
  // print the temperature:
  lcd.print(Temperature2);
  // set the cursor to column 12 of line 3
  lcd.setCursor(12, 3);
  int Temperature3=analogRead(A3)/2.05;
  // print the temperature:
  lcd.print(Temperature3);
}
```

Here is a picture of the four line LCD displaying an analog meter like display.

Here is the sketch for the four line analog meter simulation.

```
/*****************************************
LiquidCrystal Library - Analog meter
Demonstrates the use a 20x4 LCD display.
This sketch displays a simulated analog meter to the LCD

 The circuit:
* LCD RS pin to digital pin 12
* LCD En pin to digital pin 11
* LCD D4 pin to digital pin 5
* LCD D5 pin to digital pin 4
* LCD D6 pin to digital pin 3
* LCD D7 pin to digital pin 2
* LCD R/W pin to ground
* Variable resistor wiper to LCD VO pin (pin 3)

Adapted from code written by David A. Mellis
Modified by Limor Fried, Tom Igoe, Bob Davis
*****************************************/

// include the library code:
#include <LiquidCrystal.h>

// initialize the library with interface pin #'s
LiquidCrystal lcd(12, 11, 5, 4, 3, 2);

byte line1[8]= {
  B10000, B10000, B10000, B10000,
  B10000, B10000, B10000, B10000
};
byte line2[8]= {
  B01000, B01000, B01000, B01000,
  B01000, B01000, B01000, B01000
};
byte line3[8]= {
  B00100, B00100, B00100, B00100,
  B00100, B00100, B00100, B00100
};
byte line4[8]= {
  B00010, B00010, B00010, B00010,
```

```
  B00010, B00010, B00010, B00010
};
byte line5[8]= {
  B00001, B00001, B00001, B00001,
  B00001, B00001, B00001, B00001
};

void setup() {
  // set up the LCD's number of columns and rows:
  lcd.begin(20, 4);
  // Print the text to the LCD.
  lcd.createChar(1, line1);
  lcd.createChar(2, line2);
  lcd.createChar(3, line3);
  lcd.createChar(4, line4);
  lcd.createChar(5, line5);
}

void loop() {
  lcd.clear();
  lcd.setCursor(0,0);
  lcd.print("Volts A0:");
  int Voltage=analogRead(A0)/2.05;
  // print the voltage:
  lcd.setCursor(11, 0);
  lcd.print(Voltage);
  // overwrite first digit and add a .
  lcd.setCursor(10, 0);
  if (Voltage >= 100) lcd.print(Voltage/100);
  lcd.setCursor(11, 0);
  lcd.print(".");

  if (Voltage < 100) lcd.print(Voltage);
  // print the Meter:
  int metervolts=Voltage/3;
  lcd.setCursor(metervolts/10, 1);
  if (metervolts%10/2 == 0) lcd.write(1);
  if (metervolts%10/2 == 1) lcd.write(2);
  if (metervolts%10/2 == 2) lcd.write(3);
  if (metervolts%10/2 == 3) lcd.write(4);
  if (metervolts%10/2 == 4) lcd.write(5);
```

```
  lcd.setCursor(0,2);
  lcd.print("Volts A1:");
  int Voltage1=analogRead(A1)/2.05;
  // print the voltage:
  lcd.setCursor(11, 2);
  lcd.print(Voltage1);
  // overwrite first digit and add a .
  lcd.setCursor(10, 2);
  if (Voltage1 >= 100) lcd.print(Voltage1/100);
  lcd.setCursor(11, 2);
  lcd.print(".");
  if (Voltage1 < 100) lcd.print(Voltage1);
  // print the Meter:
  int metervolts1=Voltage1/3;
  lcd.setCursor(metervolts1/10, 3);
  if (metervolts1%10/2 == 0) lcd.write(1);
  if (metervolts1%10/2 == 1) lcd.write(2);
  if (metervolts1%10/2 == 2) lcd.write(3);
  if (metervolts1%10/2 == 3) lcd.write(4);
  if (metervolts1%10/2 == 4) lcd.write(5);
  delay(500);
}
```

Chapter 7

Low Resolution Graphic LCD

48 x 84 Nokia 5110

Our next project will be to set up our first graphics type of LCD. Graphics LCD's are very diverse. Unlike text based LCD's, they do not use the same controller, or the same interface chip. Even the pin connection can vary widely. On the Nokia LCD displays look for a pin with a square pad around it. The square pan usually designates pin number one.

The drivers for graphics LCD's are not included in the Arduino library. You have to download the compressed drivers for each graphics LCD and then extract or unzip the drivers into the "Arduino\libraries" directory. For the Nokia LCD, the compressed driver file is called "PCD8544_2010.zip".

The next picture is a screen capture is what the Arduino\Libraries directory looks like with the PCD8544 or Nokia LCD drivers unzipped into it.

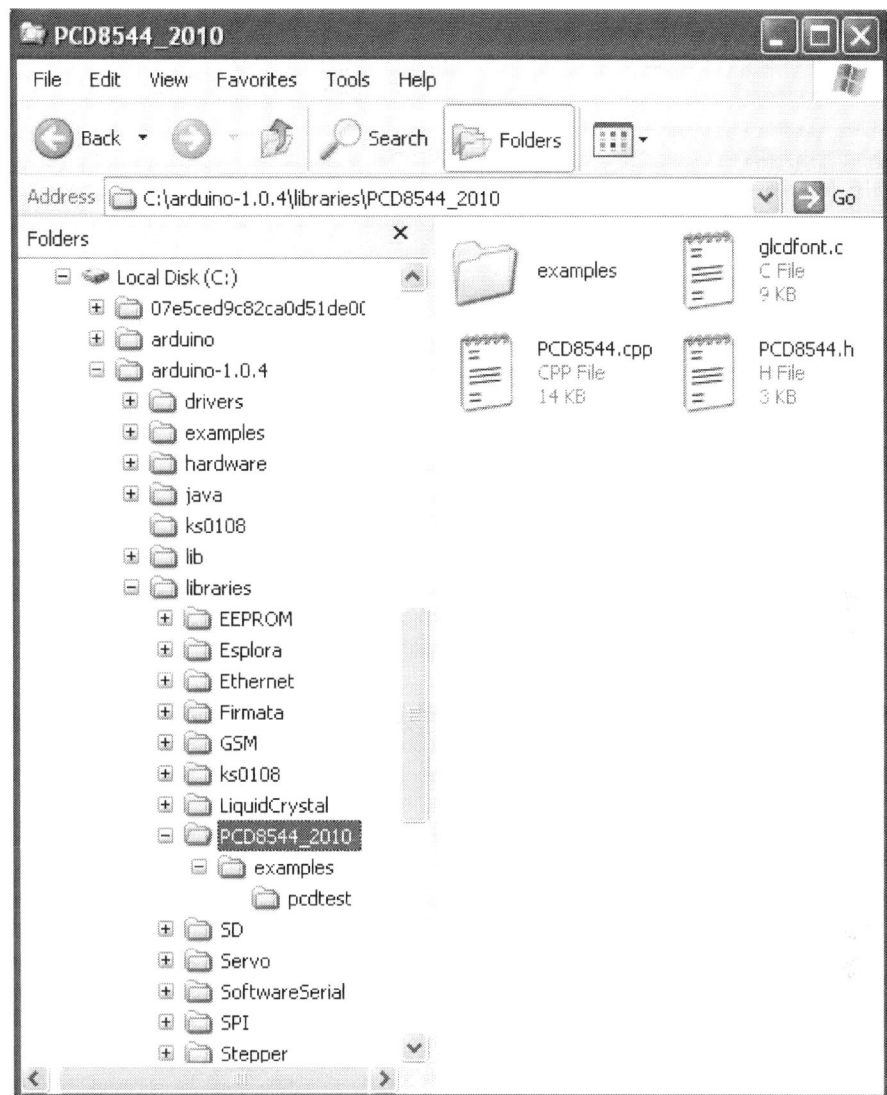

Once the files are properly extracted into the Arduino library directory, they should also show up in the Arduino interface program like what you will see in the next picture.

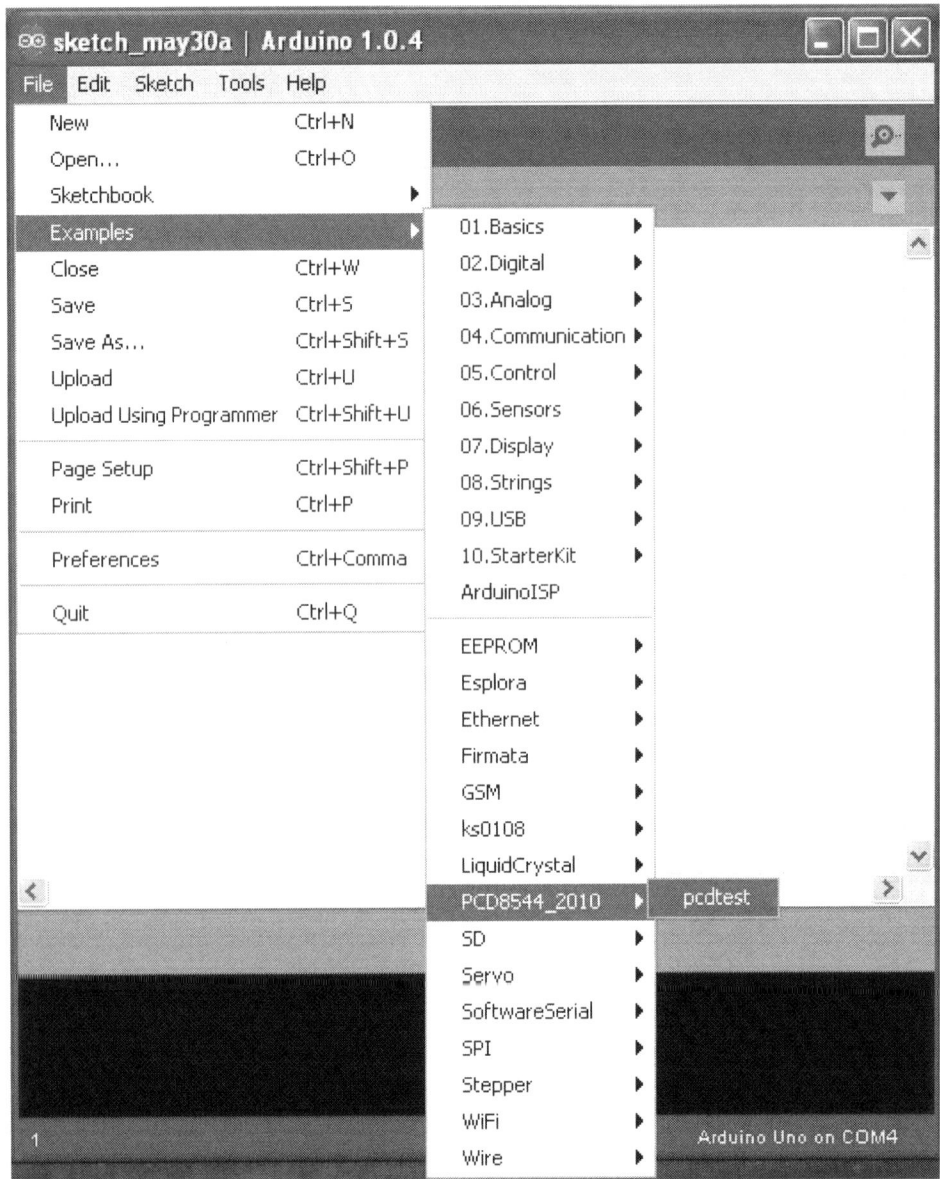

Once all of the drivers are correctly installed, then you can plug in the Arduino interfaced with the Nokia LCD, and hopefully get it to work correctly.

Here are some of the commands that the PCD8544 library adds. To get more of the graphic commands you will also need to load a graphics library.

nokia.command(command); nokia.clear();
nokia.display(); nokia.drawchar(x,y,'A');
nokia.drawcircle(x,y,r,color); nokia.drawrect(x1,y1,x2,y2,color);
nokia.drawstring(x, y, "string"); nokia.init();
nokia.setPixel(x, y, color);

Up next is the schematic diagram of how to connect the Nokia LCD to the Arduino UNO. I used five 1K resistors as "level shifters" because the logic level for the Nokia display is 3.3 volts, not the usual 5 volts. Some say these level shifter resistors are not needed, but I used them just in case.

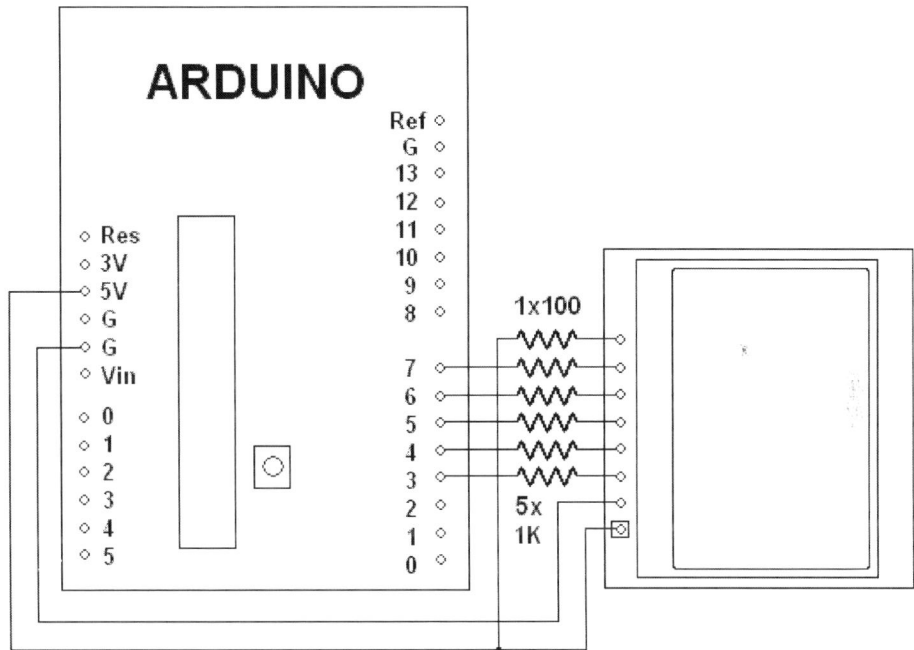

The next picture shows my quick six line text testing program running on the Nokia LCD screen.

37

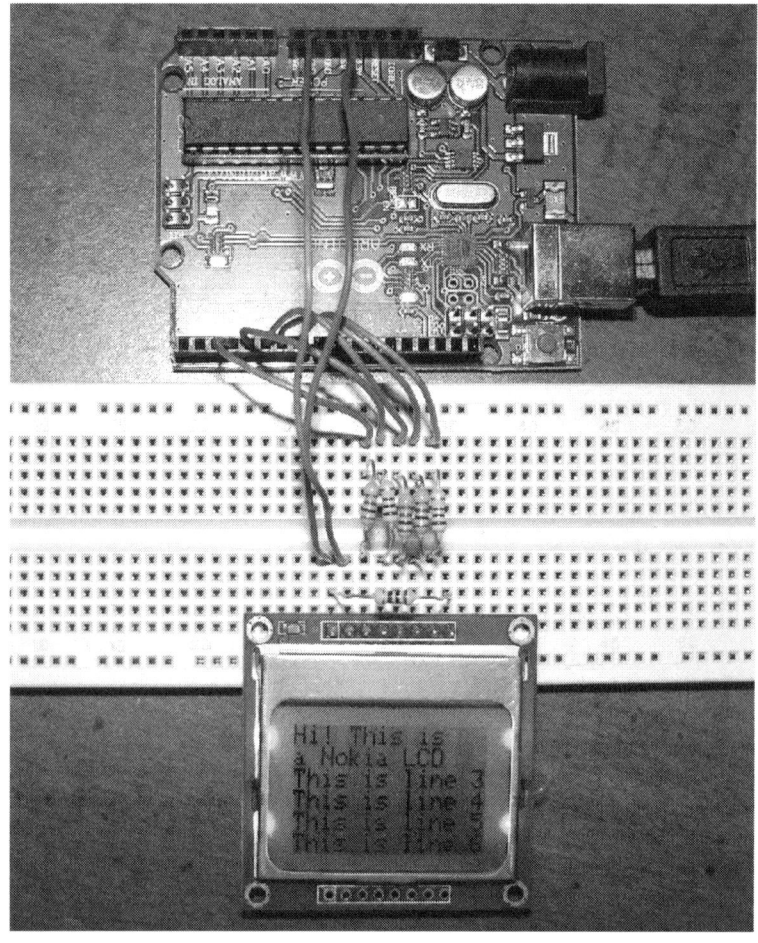

Here is the code for the quick test sketch. You can use this sketch to see if the LCD is working properly.

```
/*********************************************
  Title:    Nokia Test Display
  Purpose:   Use a Nokia LCD screen with the arduino UNO
  Created by: Bob Davis
  Note:      This code uses the Adafruit PDC8544 LCD library
*********************************************/
#include "PCD8544.h"
//Define the pins you are connecting the LCD too.
//PCD8544(SCLK, DIN/MOSI, D/C, SCE/CS, RST);
PCD8544 nokia = PCD8544(7,6,5,3,4);

void setup(void){
```

```
  nokia.init(); //Initialize lcd
  // set display to normal mode
  nokia.command(PCD8544_DISPLAYCONTROL | PCD8544_DISPLAYNORMAL);
  nokia.clear(); //clears the display
}
void loop(void){
  // Write a string to the LCD buffer use this format.
  // nokia.drawstring(x, y, "string"); x = 0-84, y = 0-48
  nokia.drawstring(0, 0, "Hi! This is   a Nokia LCD");
  nokia.drawstring(0, 16, "This is line 3");
  nokia.drawstring(0, 24, "This is line 4");
  nokia.drawstring(0, 32, "This is line 5");
  nokia.drawstring(0, 40, "This is line 6");

  //Now push the buffer to the LCD for display
  nokia.display();
}
```

The next sketch will show the contents of all six of the analog input ports on one screen.

Below is a picture of that sketch running with the Nokia LCD display.

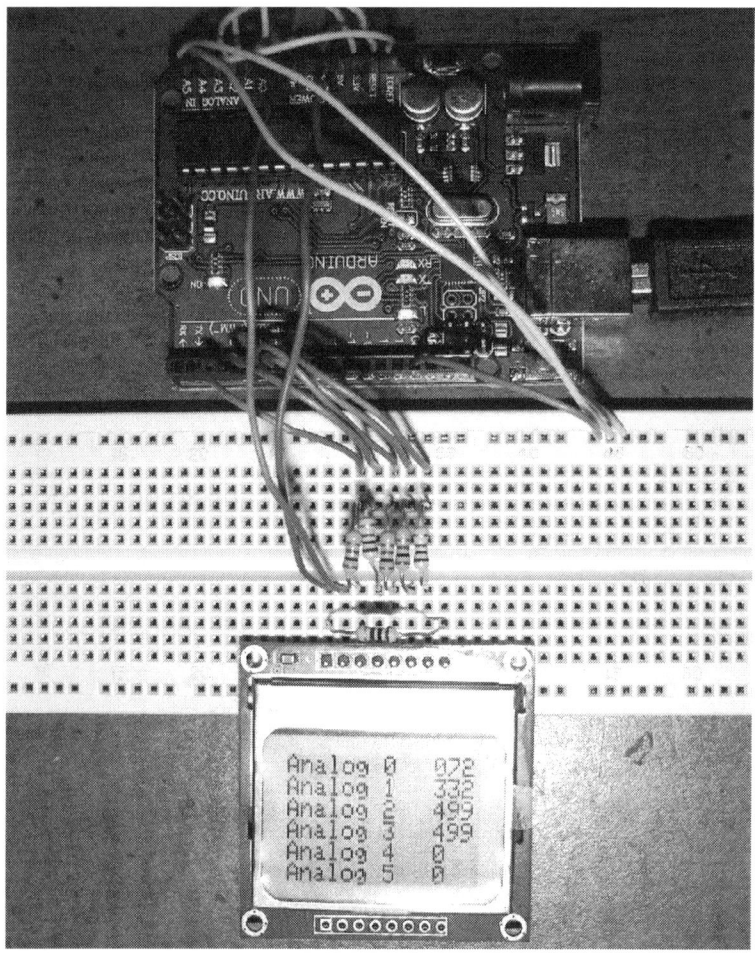

Here is the sketch code that shows the six analog inputs and displays them in millivolts.

```
/*********************************************
  Title:    Nokia Read Analog
  Purpose:   Use a Nokia LCD screen with the Arduino UNO
  Created by: Bob Davis
  Note:     This code uses the Adafruit PDC8544 LCD library
*********************************************/
#include "PCD8544.h"

//Define the pins you are connecting the LCD too.
//PCD8544(SCLK, DIN/MOSI, D/C, SCE/CS, RST);
PCD8544 nokia = PCD8544(7,6,5,3,4);
```

40

```
void setup(void){
 nokia.init(); //Initialize lcd
 // set display to normal mode
 nokia.command(PCD8544_DISPLAYCONTROL |
PCD8544_DISPLAYNORMAL);
 nokia.clear(); //clears the display
}

void loop(void){
char buf[12];
// Write a string to the LCD buffer use this format.
// nokia.drawstring(x, y, "string"); x = 0-84, y = 0-48
 nokia.drawstring(0, 0, "Analog 0");
  int analog0 = analogRead(A0)/2.05;
 nokia.drawstring(64, 0, itoa(analog0, buf, 10));
 nokia.drawstring(0, 8, "Analog 1");
 int analog1 = analogRead(A1)/2.05;
 nokia.drawstring(64, 8, itoa(analog1, buf, 10));
 nokia.drawstring(0, 16, "Analog 2");
 int analog2 = analogRead(A2)/2.05;
 nokia.drawstring(64, 16, itoa(analog2, buf, 10));
 nokia.drawstring(0, 24, "Analog 3");
 int analog3 = analogRead(A3)/2.05;
 nokia.drawstring(64, 24, itoa(analog3, buf, 10));
 nokia.drawstring(0, 32, "Analog 4");
 int analog4 = analogRead(A4)/2.05;
 nokia.drawstring(64, 32, itoa(analog4, buf, 10));
 nokia.drawstring(0, 40, "Analog 5");
 int analog5 = analogRead(A5)/2.05;
 nokia.drawstring(64, 40, itoa(analog5, buf, 10));

 //Now push the buffer to the LCD for display
 nokia.display();
}
```

When I was a teenager, I tried many times, and failed to successfully build an oscilloscope. I wanted an oscilloscope so that I could see various waveforms in order to troubleshoot some electronics problems. Back then, building an oscilloscope required a large vacuum tube for the display, some deflection amplifier tubes, and high voltage power supplies.

Now, a miniature oscilloscope can be created using an Arduino and a Nokia graphics LCD display.

For the oscilloscope schematic I added an input protection circuit for the analog input. It consists of a five volt zener diode to ground and a 100 ohm resistor in series. For a 10 x or 50 volt maximum input signal, you could also add a 900 K resistor in series with the input and a 100 K resistor to ground. A switch could then be added to select either the 1X or the 10X inputs. A .47 uF 100 volt capacitor can also be switched into series with the input. That switch then would select either AC with the capacitor or DC without it for the input coupling.

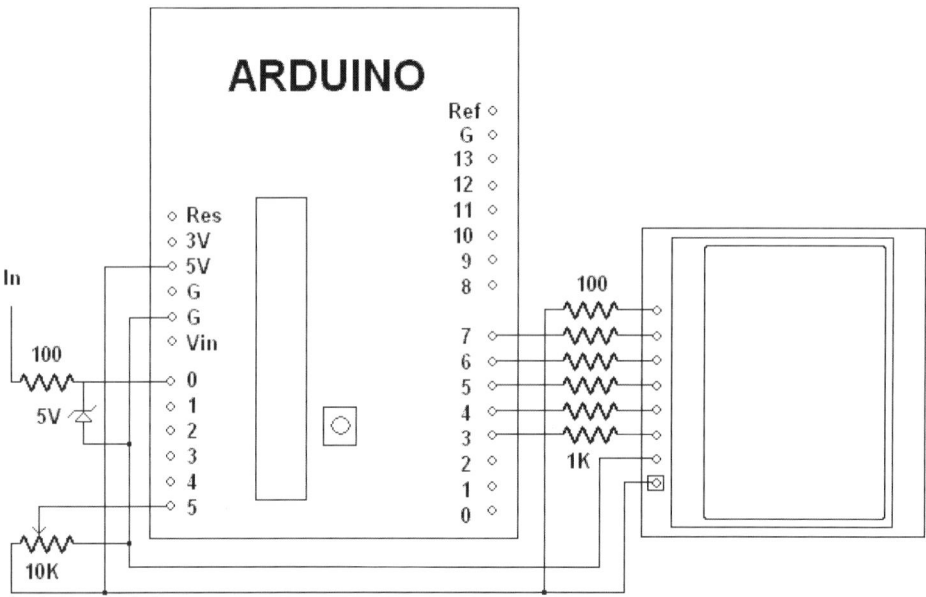

The next picture shows this oscilloscope in operation. It has no "trigger" circuit or software so the waveforms come up at different positions most of the time. It is also not very fast so it can only display lower frequencies.

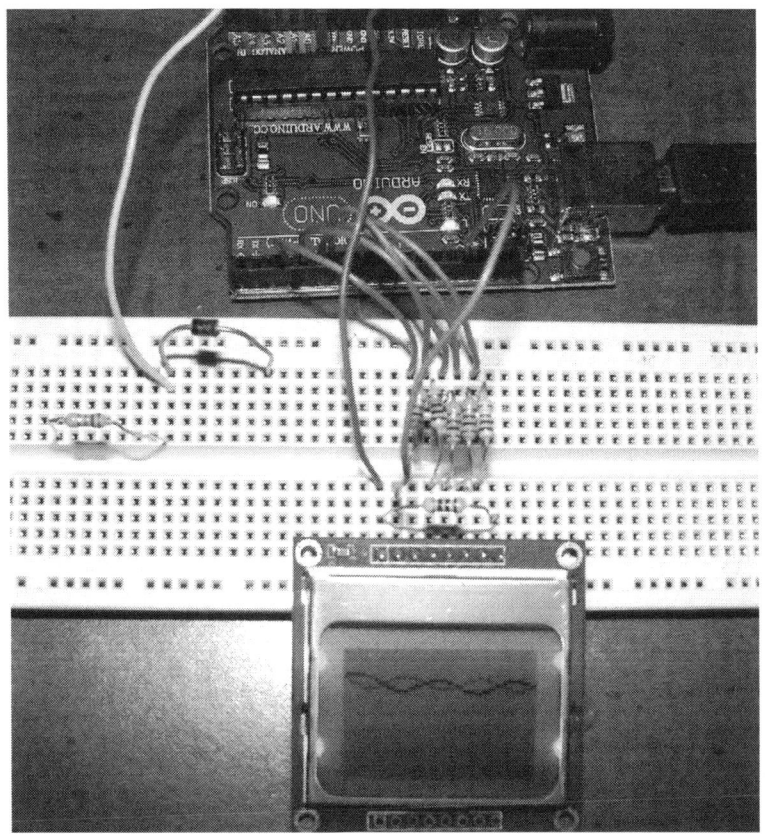

Here is a sketch code for a simple Arduino based oscilloscope.

```
/*********************************************
  Title:     Nokia Simple Arduino Oscilloscope
  Purpose:   Use a Nokia LCD screen with the arduino UNO
  Created by: Fileark see Fileark.com, Edited by Bob Davis
  Note:      This code uses the Adafruit PDC8544 LCD library
*********************************************/
#include "PCD8544.h"
PCD8544 nokia = PCD8544(7,6,5,3,4);
void setup(void) {
  nokia.init();
}

void loop() {
  int samplerate = analogRead(A5)/10;  // range of 0 to 100
  nokia.clear();        // clear the LCD
```

```
  for(int ypos = 0; ypos < 85; ypos++){   // fill screen left to right
    nokia.setPixel(ypos, analogRead(A0)/20.5, BLACK); // 5V=full scale
    delay (samplerate);    }    // delay for sampling rate
  nokia.display();       // show the data
}
```

I noticed that there I have a sketch for an analog meter simulation for many of the text LCD's. So I will include a sketch for an analog meter simulation for the Nokia and other graphical LCD's.

This next sketch produces a simulated analog meter using the line drawing feature.

```
/*********************************************
  Title:     Nokia Analog Meter
  Purpose:   Use a Nokia LCD screen with the arduino UNO
  Created by: Bob Davis
  Note:      This code uses the Adafruit PDC8544 LCD library
*********************************************/
#include "PCD8544.h"

//Define the pins you are connecting the LCD too.
//PCD8544(SCLK, DIN/MOSI, D/C, SCE/CS, RST);
PCD8544 nokia = PCD8544(7,6,5,3,4);

void setup(void){
  nokia.init(); //Initialize lcd
  // set display to normal mode
  nokia.command(PCD8544_DISPLAYCONTROL | PCD8544_DISPLAYNORMAL);
}

void loop(void){
  char buf[12];
  nokia.clear(); //clears the display
  // Write a string to the LCD buffer using this format.
  // nokia.drawstring(x, y, "string"); x = 0-84, y = 0-48
  nokia.drawstring(0, 0, "Analog 0");
  int analog0 = analogRead(A0)/2.05;
  nokia.drawstring(64, 0, itoa(analog0, buf, 10));
  // nokia.drawline(x1, y1, x2, y2, color);
  nokia.drawline(42, 48, analog0/5, 0, BLACK);
  //Now push the buffer to the LCD for display
  nokia.display();
}
```

Chapter 8

Medium Resolution Graphic LCD

128x64 QC12864B

When I said that graphics LCD's vary widely in their pinout, that was an understatement. When it comes to 128 by 64 graphics LCD's, they vary in their pinout, their physical size, the controller chip that is used, and even in their modes of operation.

I did not have a 128 by 64 graphics display in stock, so I had to buy one for this project. The first one arrived within a few weeks, but when I opened it, there was a 40 pin ribbon cable in the box instead of the LCD! I quickly went to one of my favorite vendors and ordered another 128 by 64 LCD without even reading the fine print. When it arrived I connected it up and loaded the software, but all I got was garbage. So I went back and read the fine print. It is labeled QC12864B and it uses a ST7920 controller chip!

The solution is to use the universal Graphics library called "U8Glib.zip". That stands for Universal 8 bit graphics library. Once again, you need to decompress it into your Arduino\libraries directory and make sure that it shows up in the Arduino interface.

Up next is a picture of that U8Glib library properly installed.

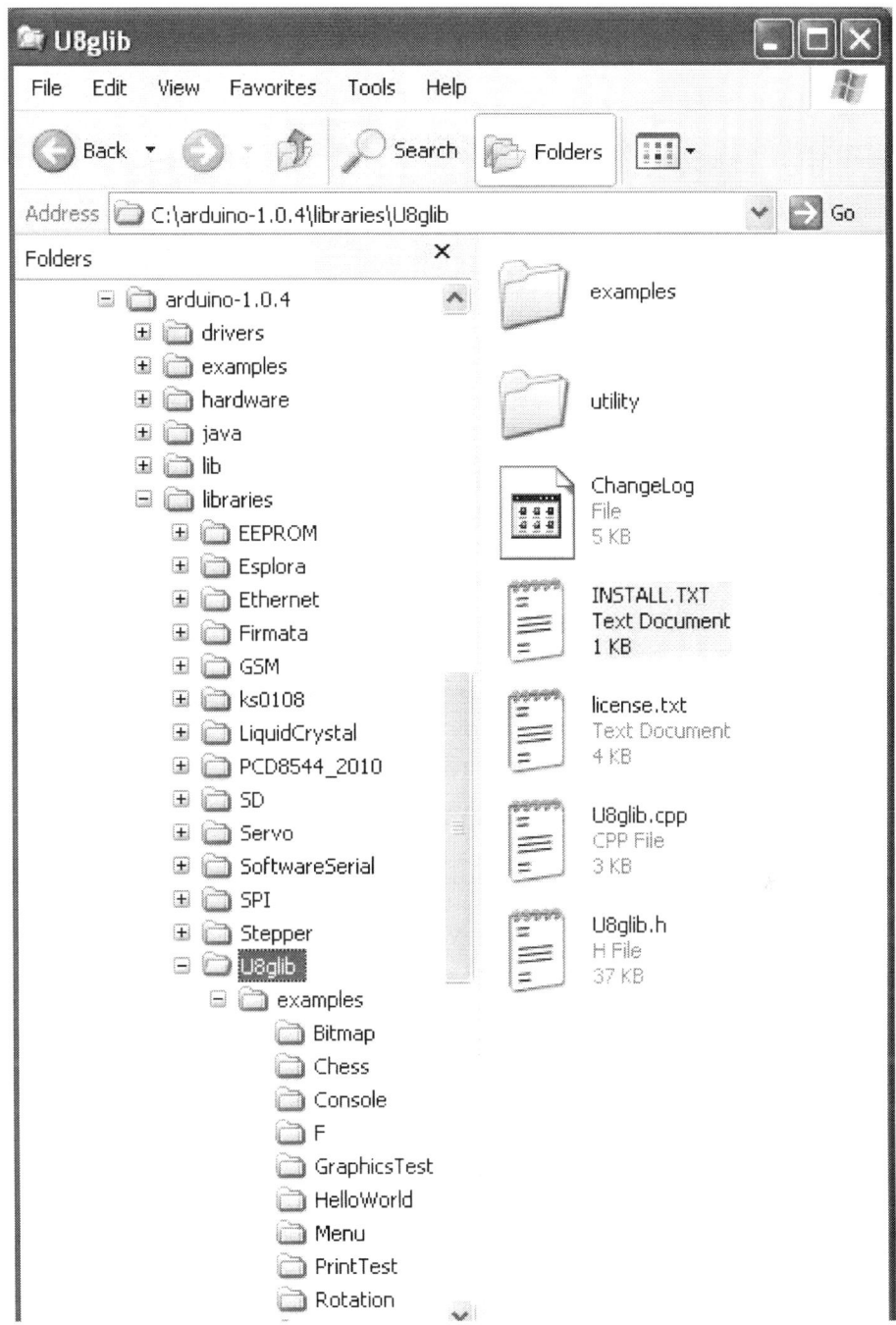

The U8G Library has a very long list of commands that it supports. Here is a quick list of the available commands:
Begin(void);
disableCursor(void);

drawBitmap, drawBitmapP(x, y, count, height, *bitmap);
drawBox(x, y, width, height);
drawCircle(x, y, radius);
drawDisc(x, y, radius);
drawFrame(x, y, width, height);
drawHLine, drawVLine(x, y, width);
drawLine(x1, y1, x2, y2);
drawPixel(x, y);
drawRBox, drawRFrame(x, y, width, height, radius);
drawStr, drawStr90, drawStr180, drawStr270(x, y, *string);
drawStrP, drawStr90P, drawStr180P, drawStr270P(x, y, *string);
drawXBM, drawXBMP(x, y, width, height, *bitmap);
enableCursor(void);
firstPage(void);
getColorIndex(void);
getFontAscent, getFontDescent, getFontLineSpacing(void);
getHeight, getMode, getWidth(void);
getStrWidth(*string);
InitSPI(*dev, sck, mosi, cs, a0, reset);
InitHWSPI(*dev, cs, a0, reset);
Init8Bit(*dev, d1, d2, d3, d4, d5, d6, d7, en, cs1, cs2, di, rw, reset);
nextPage(void);
print(. . .);
setColorIndex(Color_index);
setContrast(contrast);
setCursorColor(foreground,background);
setCursorFont(*font);
setCursorPos(x, y);
setCursorStyle(encoding);
setDefaultBackgroundColor, setDefaultForegroundColor,
setDefaultMidColor(void);
setFont(*font);
setFontLineSpacingFactor(factor)
setFontPosBaseline, setFontPosBottom, setFontPosCenter,
setFontPosTop(void);
setFontRefHeightAll, setFontRefHeightExtendedText,
setFontRefHeightText(void);
setPrintPos(x, y);
setRot90, setRot180, setRot270();
setScale2x2();
sleepOn, sleepOff(void);

undoRotation();
undoScale();

Then there is a long list of chips and graphics formats that U8Glib supports. You have to remove the "//" in front of the chip and graphics setup that you are using. Then you wire up the LCD according to the pinout that is given on the next line. This is only a partial list of the many supported chips. We will be using the ST7920 chipset.

```
//U8GLIB_NHD27OLED_BW u8g(13, 11, 10, 9);
// SPI Com: SCK = 13, MOSI = 11, CS = 10, A0 = 9
//U8GLIB_NHD31OLED_BW u8g(13, 11, 10, 9);
// SPI Com: SCK = 13, MOSI = 11, CS = 10, A0 = 9
//U8GLIB_DOGS102 u8g(13, 11, 10, 9);
// SPI Com: SCK = 13, MOSI = 11, CS = 10, A0 = 9
//U8GLIB_DOGM132 u8g(13, 11, 10, 9);
// SPI Com: SCK = 13, MOSI = 11, CS = 10, A0 = 9
//U8GLIB_DOGM128 u8g(13, 11, 10, 9);
// SPI Com: SCK = 13, MOSI = 11, CS = 10, A0 = 9
U8GLIB_ST7920_128X64_4X u8g(8, 9, 10, 11, 4, 5, 6, 7, 18, 17, 16);
// 8Bit Com: D0..D7: 8,9,10,11,4,5,6,7 en=18, di=17,rw=16
//U8GLIB_LM6059 u8g(13, 11, 10, 9);
// SPI Com: SCK = 13, MOSI = 11, CS = 10, A0 = 9
//U8GLIB_LM6063 u8g(13, 11, 10, 9);
// SPI Com: SCK = 13, MOSI = 11, CS = 10, A0 = 9
//U8GLIB_DOGXL160_BW u8g(10, 9);
// SPI Com: SCK = 13, MOSI = 11, CS = 10, A0 = 9
//U8GLIB_PCD8544 u8g(13, 11, 10, 9, 8);
// SPI Com: SCK = 13, MOSI = 11, CS = 10, A0 = 9, Reset = 8
//U8GLIB_PCF8812 u8g(13, 11, 10, 9, 8);
// SPI Com: SCK = 13, MOSI = 11, CS = 10, A0 = 9, Reset = 8
//U8GLIB_KS0108_128 u8g(8, 9, 10, 11, 4, 5, 6, 7, 18, 14, 15, 17, 16);
// 8Bit Com: D0..D7: 8,9,10,11,4,5,6,7 en=18, cs1=14,
cs2=15,di=17,rw=16
//U8GLIB_LC7981_160X80 u8g(8, 9,10,11,4,5,6,7, 18, 14, 15, 17, 16);
//8Bit Com:D0..D7:8,9,10,11,4,5,6,7 en=18, cs=14,di=15,rw=17,reset=16
//U8GLIB_ILI9325D_320x240 u8g(18,17,19,U8G_PIN_NONE,16 );
// 8Bit Com: D0..D7: 0,1,2,3,4,5,6,7 en=wr=18, cs=17, rs=19,
rd=U8G_PIN_NONE, //U8GLIB_SBN1661_122X32
u8g(8,9,10,11,4,5,6,7,14,15, 17, U8G_PIN_NONE, 16);
//8BitCom:D0..D7:8,9,10,11,4,5,6,7cs1=14,cs2=15,di=17,rw=16,reset= 16
```

//U8GLIB_SSD1306_128X64 u8g(13, 11, 10, 9);
// SW SPI Com: SCK = 13, MOSI = 11, CS = 10, A0 = 9
//U8GLIB_SSD1309_128X64 u8g(13, 11, 10, 9);
// SPI Com: SCK = 13, MOSI = 11, CS = 10, A0 = 9
//U8GLIB_NHD_C12864 u8g(13, 11, 10, 9, 8);
// SPI Com: SCK = 13, MOSI = 11, CS = 10, A0 = 9, RST = 8
//U8GLIB_NHD_C12832 u8g(13, 11, 10, 9, 8);
// SPI Com: SCK = 13, MOSI = 11, CS = 10, A0 = 9, RST = 8
//U8GLIB_T6963_240X128 u8g(8,9,10,11,4,5,6,7,14,15, 17, 18, 16);
// 8Bit Com: D0..D7: 8,9,10,11,4,5,6,7,cs=14,a0=15,wr=17,rd=18, reset=16

Below is a picture of the pinout sheet for the QC12864B 128 by 64 graphics LCD.

ITEM	SYMBOL	LEVEL	FUNCTIONS
1	VSS	0V	Power Ground
2	VDD	+5V	Power supply for logic
3	V0	—	Contrast adjust
4	RS(CS)	H/L	H:data L:command
5	RW/(SID)	H/L	H:read L:write
6	E/(SCLK)	H.H→L	Enable signal
7-14	DB0-DB7	H/L	Data Bus
15	PSB	H/L	H:Paraller mode L:serial mode
16	NC		No connection
17	/REST	L	Reset signal
18	VOUT	—	Output LCD voltage
19	LEDA	+5V	Power supply for LED backlight
20	LEDK	0V	

If you think that chart is confusing, then you are not alone. On top of that, we have the connections to the Arduino UNO to consider. Here is the pinout showing the Arduino connections and written in plain English.

```
Pin    Symbol      Arduino      Function
----   ---------   ----------   ----------
1      Vss         GND          Ground
2      Vdd         5V           Five Volts
3      Vo                       Contrast variable resistor wiper
```

4	RS/CS	17 - A3	Register Select
			High is data, low is a command
5	RW/SID	16 - A2	Read Write, or Serial Data
			High is read, Low is write
6	E/SCLK	18 - A4	Enable or Serial Clock
7	DB0	8	data Bus bit 0
8	DB1	9	data Bus bit 1
9	DB2	10	data Bus bit 2
10	DB3	11	data Bus bit 3
11	DB4	4	data Bus bit 4
12	DB5	5	data Bus bit 5
13	DB6	6	data Bus bit 6
14	DB7	7	data Bus bit 7
15	PSB	5V	Parallel/Serial Bus
			High is Parallel, Low is Serial
16	NC		Not Connected – a really easy one
17	RST	5V	Reset Signal – Low is Reset
18	Vout		Voltage out to the contrast variable resistor
19	LED+	5V	Anode to the LED backlight
20	LED	GND	Ground for the LED backlight

Once the support software is installed you can wire it up and plug it in. My LCD still produced garbage, but it was because I had some data wires swapped. I used all white jumpers. Some people recommend using different color jumper wires for each bit so that they do not get crossed by accident.

Up next is the schematic diagram for the QC12864B in parallel eight bit operation.

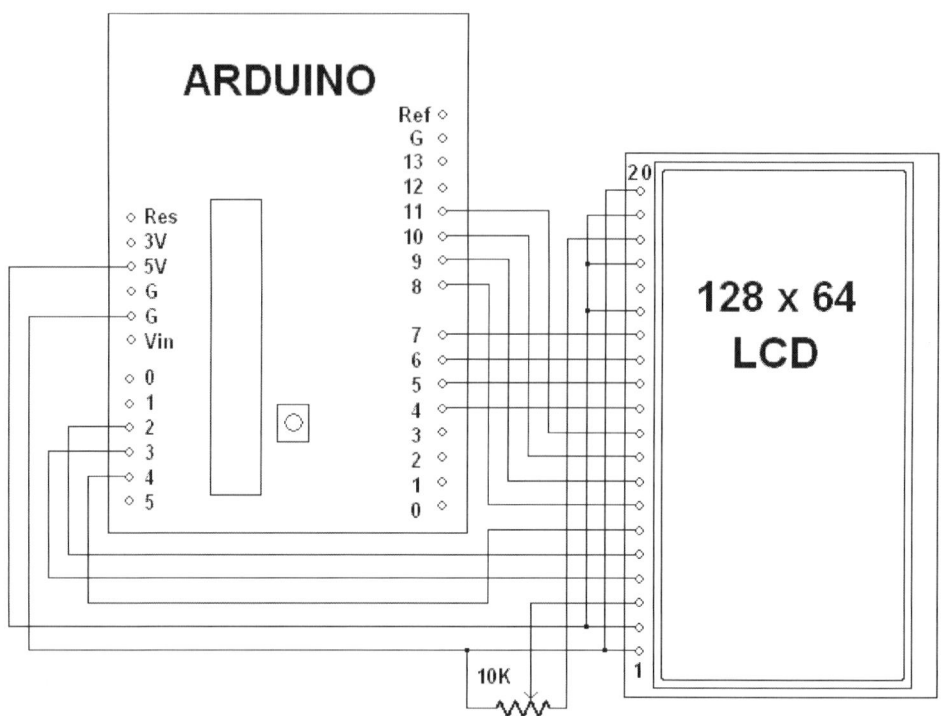

The first project we will build with this LCD produces a simulated analog meter. A variable resistor can be used for test purposes. The analog input is connected to A0. You can also use the audio output of your computer, instead of the variable resistor, to make the meter dance with the music.

Up next is a picture of the simulated analog meter sketch in operation with the QC12864B.

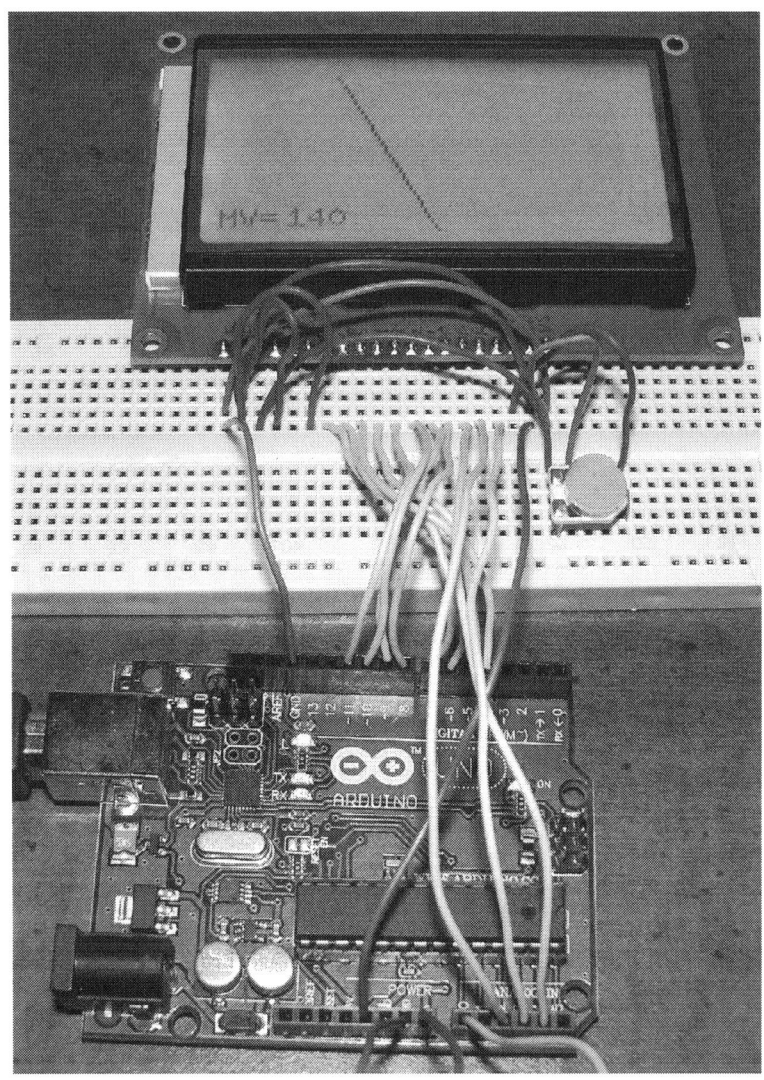

Here is a sketch to create the simulated analog meter.

/**********************************
128 by 64 LCD Analog meter simulation
By Bob Davis
Uses Universal 8bit Graphics Library, http://code.google.com/p/u8glib/

Copyright (c) 2012, olikraus@gmail.com   All rights reserved.

***/
#include "U8glib.h"

```
// 8Bit Com: D0..D7: 8,9,10,11,4,5,6,7 en=18, di=17,rw=16
U8GLIB_ST7920_128X64_4X u8g(8, 9, 10, 11, 4, 5, 6, 7, 18, 17, 16);

void u8g_prepare(void) {
  u8g.setFont(u8g_font_6x10);
  u8g.setFontRefHeightExtendedText();
  u8g.setDefaultForegroundColor();
  u8g.setFontPosTop();
}

//uint8_t draw_state = 0;
void draw(void) {
  char buf[12];
  u8g_prepare();
  int Analog0=analogRead(A0)/2.05;
  u8g.drawStr( 0, 54, "MV= ");
  u8g.drawStr( 20, 54, itoa(Analog0, buf, 10));
  u8g.drawLine(64, 64, Analog0/4, 0);
}

void setup(void) {
  // assign default color value
  if ( u8g.getMode() == U8G_MODE_R3G3B2 )
    u8g.setColorIndex(255);     // RGB=white
  else if ( u8g.getMode() == U8G_MODE_GRAY2BIT )
    u8g.setColorIndex(3);       // max intensity
  else if ( u8g.getMode() == U8G_MODE_BW )
    u8g.setColorIndex(1);       // pixel on, black
}

void loop(void) {
  // picture loop
  u8g.firstPage();
  do { draw(); }
  while( u8g.nextPage() );
  // rebuild the picture after some delay
  delay(100);
}
```

The next project is another simple Arduino based oscilloscope. It has been greatly improved over the Nokia version. It even has a software based trigger loop so the waveform does not jump around nearly as much.

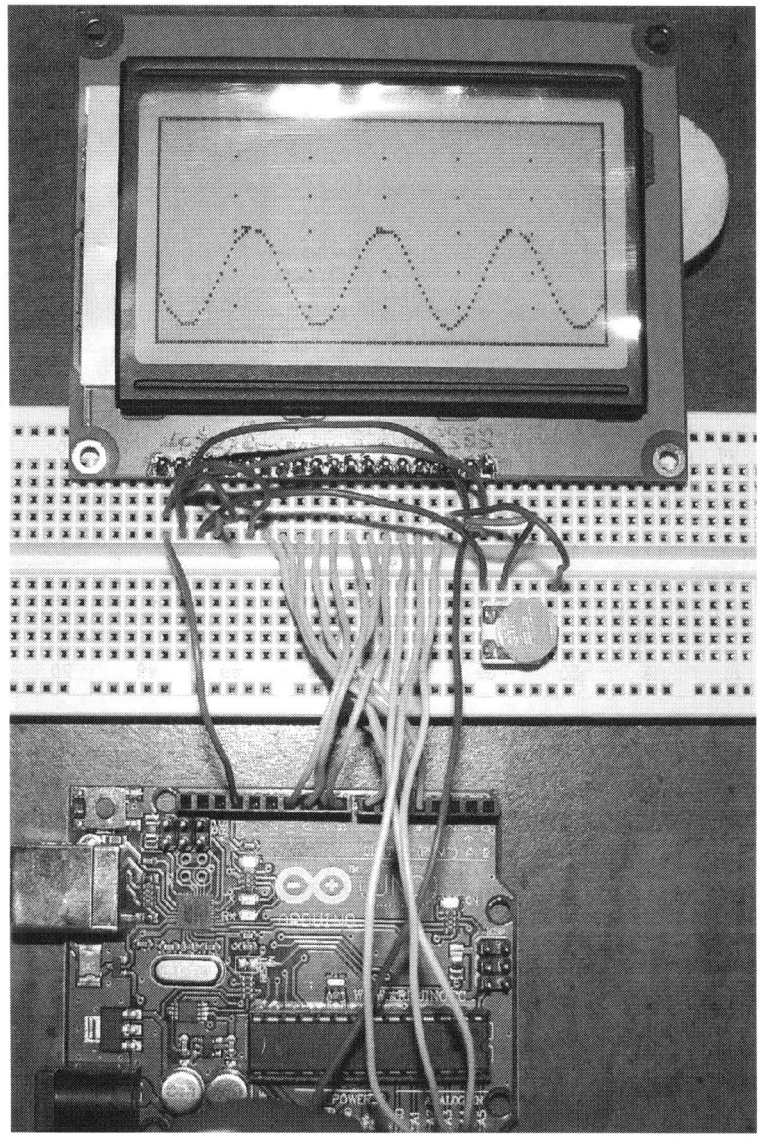

Here is a schematic diagram of the oscilloscope analog input circuit. The input circuit is not visible in the above picture because it was made with jumper wires and was located off the bottom of the picture. It is AC coupled via a .47 uF capacitor so that a variable resistor can adjust the position of the waveform on the screen. A five volt zener diode provides overload protection.

I was running a program called "Sweep Generator" on my laptop computer. It is named "sweepgen.exe". I have tested this oscilloscope design with an audio sweep range from 100 Hz to 1000 Hz. It is quite useable up to 1 KHz, but it is not very usable above that frequency. The display starts looking like a lot of noise above 1 KHz.

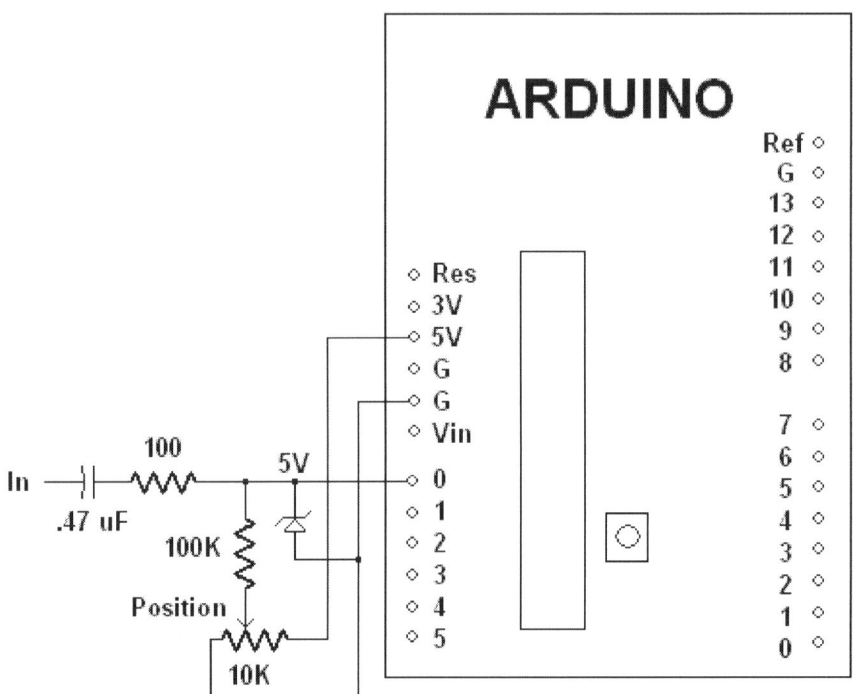

Here is the improved oscilloscope sketch.

```
/*********************************
128 by 64 LCD Oscilloscope
By Bob Davis
Uses Universal 8bit Graphics Library, http://code.google.com/p/u8glib/
  Copyright (c) 2012, olikraus@gmail.com All rights reserved.
*********************************/
#include "U8glib.h"
// 8Bit Com: D0..D7: 8,9,10,11,4,5,6,7 en=18, di=17,rw=16
U8GLIB_ST7920_128X64_4X u8g(8, 9, 10, 11, 4, 5, 6, 7, 18, 17, 16);

int Sample[128];
int Input=0;
int OldInput=0;
```

```
int rate=0;

void u8g_prepare(void) {
  u8g.setFont(u8g_font_6x10);
  u8g.setFontRefHeightExtendedText();
  u8g.setDefaultForegroundColor();
  u8g.setFontPosTop();
}
void DrawMarkers(void) {
  u8g.drawFrame (0,0,128,64);
  u8g.drawPixel (22,16);
  u8g.drawPixel (43,16);
  u8g.drawPixel (64,16);
  u8g.drawPixel (85,16);
  u8g.drawPixel (106,16);
  u8g.drawPixel (22,32);
  u8g.drawPixel (43,32);
  u8g.drawPixel (64,32);
  u8g.drawPixel (85,32);
  u8g.drawPixel (106,32);
  u8g.drawPixel (22,48);
  u8g.drawPixel (43,48);
  u8g.drawPixel (64,48);
  u8g.drawPixel (85,48);
  u8g.drawPixel (106,48);
}

void sample_data(void){
// wait for a trigger of a positive going input
//  Input=analogRead(A0);
  while (Input < OldInput){
    OldInput=analogRead(A0);
    Input=analogRead(A0);
  }
// collect the analog data into an array
// do not do division by 10.24 here, it makes it slower!
  for(int xpos=0; xpos<128; xpos++) {
    Sample[xpos]=analogRead(A0);
  }
}
```

```
void draw(void) {
  u8g_prepare();
  DrawMarkers();
// display the collected analog data from array
// Sample/10.24 because 1024 becomes 100 = 5 volts
  for(int xpos=1; xpos<128; xpos++) {
    u8g.drawLine (xpos, Sample[xpos]/10.24, xpos+1,
Sample[xpos+1]/10.24);
  }
}

void setup(void) {
  // assign default color value
  if ( u8g.getMode() == U8G_MODE_R3G3B2 )
    u8g.setColorIndex(255);     // RGB=white
  else if ( u8g.getMode() == U8G_MODE_GRAY2BIT )
    u8g.setColorIndex(3);       // max intensity
  else if ( u8g.getMode() == U8G_MODE_BW )
    u8g.setColorIndex(1);       // pixel on, black
}
void loop(void) {
// Set sample speed according to switch on A1
  // picture loop
  sample_data();
  u8g.firstPage();
  do { draw(); }
  while( u8g.nextPage() );
  // rebuild the picture after some delay
  delay(100);
}
```

Up next is a picture of the two channel fast logic analyzer. The schematic is the same as the one for the oscilloscope, but the position control and the 100 K resistor are removed as they are not needed. The sample rate is around two million samples per second, as it uses the Parallel Input command "PINC".

The waveform being displayed is of a 15,000 Hz two phase audio signal from the computer program "Audio SweepGen" running on my computer.

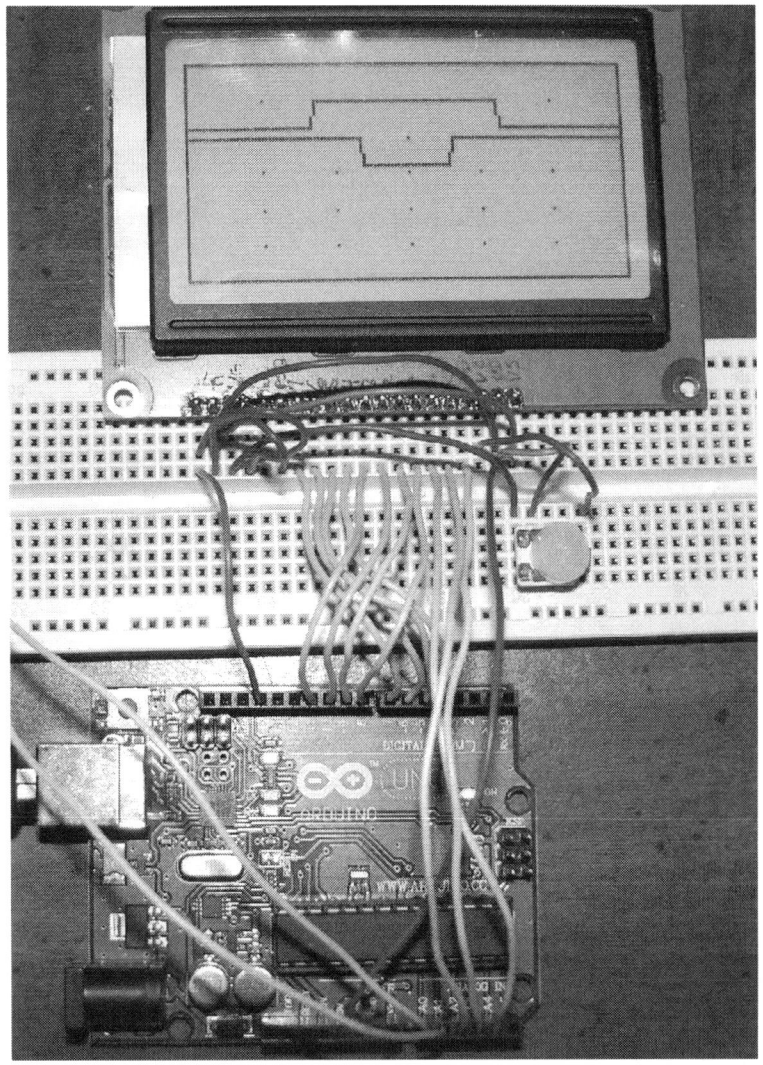

Here is the sketch for a two channel logic analyzer.

/*********************************
128 by 64 LCD Logic Analyzer
By Bob Davis
Uses Universal 8bit Graphics Library, http://code.google.com/p/u8glib/
  Copyright (c) 2012, olikraus@gmail.com   All rights reserved.

*********************************/
#include "U8glib.h"

// 8Bit Com: D0..D7: 8,9,10,11,4,5,6,7 en=18, di=17,rw=16

```
U8GLIB_ST7920_128X64_4X u8g(8, 9, 10, 11, 4, 5, 6, 7, 18, 17, 16);

int Sample[128];
int Sample0[128];
int Sample1[128];
int Input=0;
int OldInput=0;

void u8g_prepare(void) {
  u8g.setFont(u8g_font_6x10);
  u8g.setFontRefHeightExtendedText();
  u8g.setDefaultForegroundColor();
  u8g.setFontPosTop();
}
void DrawMarkers(void) {
  u8g.drawFrame (0,0,128,64);
  u8g.drawPixel (22,16);
  u8g.drawPixel (43,16);
  u8g.drawPixel (64,16);
  u8g.drawPixel (85,16);
  u8g.drawPixel (106,16);
  u8g.drawPixel (22,32);
  u8g.drawPixel (43,32);
  u8g.drawPixel (64,32);
  u8g.drawPixel (85,32);
  u8g.drawPixel (106,32);
  u8g.drawPixel (22,48);
  u8g.drawPixel (43,48);
  u8g.drawPixel (64,48);
  u8g.drawPixel (85,48);
  u8g.drawPixel (106,48);
}

void draw(void) {
  u8g_prepare();
  DrawMarkers();
// wait for a trigger of a positive going input
//  OldInput=digitalRead(A0);
  Input=digitalRead(A0);
  while (Input != 1){
    Input=digitalRead(A0);
```

```
  }
// collect the analog data into an array
  for(int xpos=0; xpos<128; xpos++) {
    Sample[xpos]=PINC;
  }
// Process the data to get it ready to display
  for(int xpos=0; xpos<128; xpos++) {
    Sample0[xpos]=(((Sample[xpos]&B00000001)*8)+11);
    Sample1[xpos]=(((Sample[xpos]&B00000010)*4)+22);
  }

// display the collected analog data from array
  for(int xpos=0; xpos<128; xpos++) {
    u8g.drawLine (xpos, Sample0[xpos], xpos+1, Sample0[xpos+1]);
    u8g.drawLine (xpos, Sample1[xpos], xpos+1, Sample1[xpos+1]);
  }
}

void setup(void) {
  // assign default color value
  pinMode(A0, INPUT);
  pinMode(A1, INPUT);
  if ( u8g.getMode() == U8G_MODE_R3G3B2 )
    u8g.setColorIndex(255);     // RGB=white
  else if ( u8g.getMode() == U8G_MODE_GRAY2BIT )
    u8g.setColorIndex(3);       // max intensity
  else if ( u8g.getMode() == U8G_MODE_BW )
    u8g.setColorIndex(1);       // pixel on, black
}

void loop(void) {
// picture loop
//  u8g.firstPage();
  do { draw(); }
  while( u8g.nextPage() );
  // rebuild the picture after some delay
  delay(100);
}
```

The next project is a six channel logic analyzer. To make all six of the analog inputs available we need to move the control signals for the LCD to other pins. D1, D2 and D3 are all available with this LCD so I moved the control signals there. Here is an updated schematic with the control pins moved. This change leaves all six analog inputs free to be used as logic inputs.

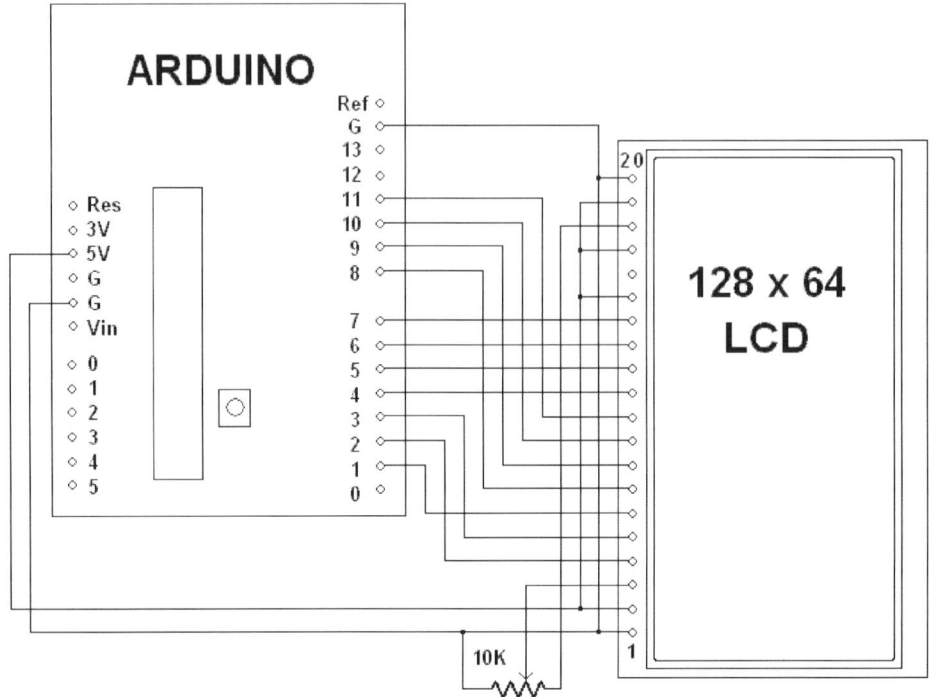

Coming up is a picture of the six channel logic analyzer in use. It is monitoring the 100 KC to 1 MC outputs of a 74LS390 dual divide by 10 IC with a 10 Mc Clock input.

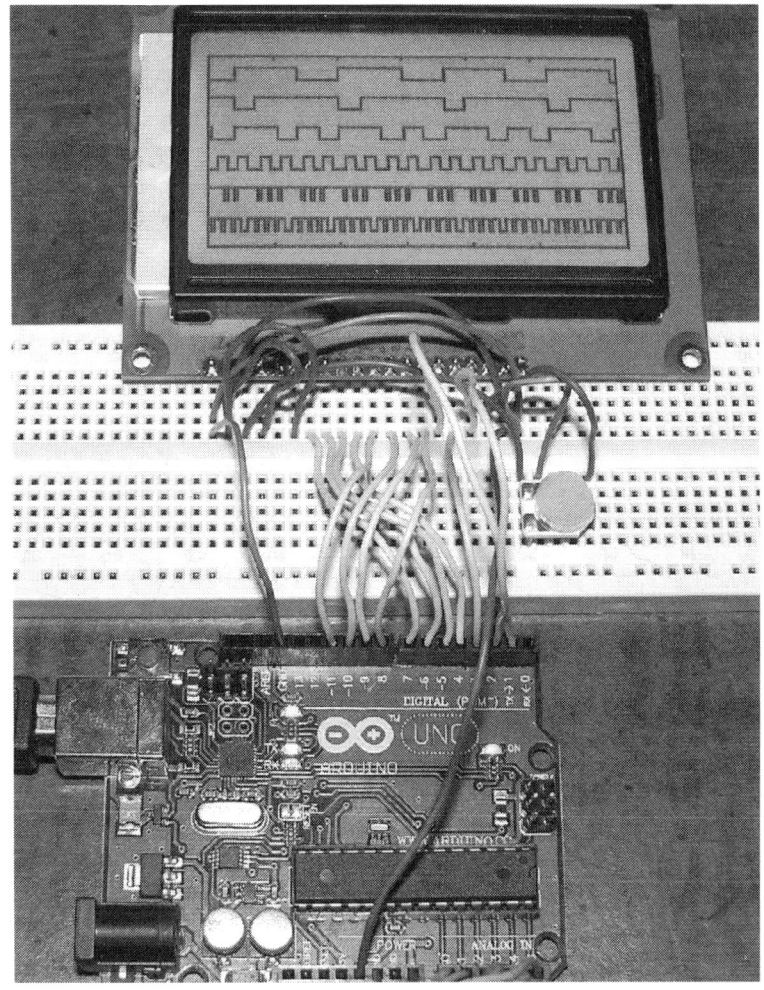

Here is the sketch code to make the six channel logic analyzer work.

```
/*********************************
128 by 64 six channel LCD Logic Analyzer
By Bob Davis
Uses Universal 8bit Graphics Library, http://code.google.com/p/u8glib/

  Copyright (c) 2012, olikraus@gmail.com   All rights reserved.
*********************************/
#include "U8glib.h"

// 8Bit Com: D0..D7: 8,9,10,11,4,5,6,7 en=18, di=17,rw=16
//U8GLIB_ST7920_128X64_4X u8g(8, 9, 10, 11, 4, 5, 6, 7, 18, 17, 16);
//  **** NOTE **** I Moved the three control pins !!!
```

```
U8GLIB_ST7920_128X64_4X u8g(8, 9, 10, 11, 4, 5, 6, 7, 1, 2, 3);

int Sample[128];
int Input=0;
int OldInput=0;

void u8g_prepare(void) {
  u8g.setFont(u8g_font_6x10);
  u8g.setFontRefHeightExtendedText();
  u8g.setDefaultForegroundColor();
  u8g.setFontPosTop();
}

void DrawMarkers(void) {
  u8g.drawFrame (0,0,128,64);
  u8g.drawPixel (20,1);
  u8g.drawPixel (40,1);
  u8g.drawPixel (60,1);
  u8g.drawPixel (80,1);
  u8g.drawPixel (100,1);
  u8g.drawPixel (20,62);
  u8g.drawPixel (40,62);
  u8g.drawPixel (60,62);
  u8g.drawPixel (80,62);
  u8g.drawPixel (100,62);
}

void draw(void) {
  u8g_prepare();
  DrawMarkers();
// wait for a trigger of a positive going input
  Input=digitalRead(A0);
  while (Input != 1){
    Input=digitalRead(A0);
  }

// collect the analog data into an array
  for(int xpos=0; xpos<128; xpos++) {
    Sample[xpos]=PINC;
  }
```

```
// display the collected analog data from array
  for(int xpos=0; xpos<128; xpos++) {
    u8g.drawLine (xpos, ((Sample[xpos]&B00000001)*4)+4, xpos+1, ((Sample[xpos+1]&B00000001)*4)+4);
    u8g.drawLine (xpos, ((Sample[xpos]&B00000010)*2)+14, xpos+1, ((Sample[xpos+1]&B00000010)*2)+14);
    u8g.drawLine (xpos, ((Sample[xpos]&B00000100)*1)+24, xpos+1, ((Sample[xpos+1]&B00000100)*1)+24);
    u8g.drawLine (xpos, ((Sample[xpos]&B00001000)/2)+34, xpos+1, ((Sample[xpos+1]&B00001000)/2)+34);
    u8g.drawLine (xpos, ((Sample[xpos]&B00010000)/4)+44, xpos+1, ((Sample[xpos+1]&B00010000)/4)+44);
    u8g.drawLine (xpos, ((Sample[xpos]&B00100000)/8)+54, xpos+1, ((Sample[xpos+1]&B00100000)/8)+54);
  }
}
void setup(void) {
  pinMode(A0, INPUT);
  pinMode(A1, INPUT);
  pinMode(A2, INPUT);
  pinMode(A3, INPUT);
  pinMode(A4, INPUT);
  pinMode(A5, INPUT);

  // assign default color value
  if ( u8g.getMode() == U8G_MODE_R3G3B2 )
    u8g.setColorIndex(255);     // RGB=white
  else if ( u8g.getMode() == U8G_MODE_GRAY2BIT )
    u8g.setColorIndex(3);       // max intensity
  else if ( u8g.getMode() == U8G_MODE_BW )
    u8g.setColorIndex(1);       // pixel on, black
}

void loop(void) {
// picture loop
//  u8g.firstPage();
  do { draw(); }
  while( u8g.nextPage() );
  // rebuild the picture after some delay
  delay(100);
}
```

You can take the six channel logic analyzer a step further and add an external CA3306 analog to digital converter. The CA3306 schematic is in the next chapter. You can also add speed select switches on D12 and D13 to ground. Then try out this sketch code for a five million sample per second logic analyzer or oscilloscope

```
/*********************************
128 by 64 LCD Oscilloscope ext atod speed select
By Bob Davis
Uses Universal 8bit Graphics Library, http://code.google.com/p/u8glib/
  Copyright (c) 2012, olikraus@gmail.com  All rights reserved.
*******************************************/
#include "U8glib.h"

// 8Bit Com: D0..D7: 8,9,10,11,4,5,6,7 en=18, di=17,rw=16
//U8GLIB_ST7920_128X64_4X u8g(8, 9, 10, 11, 4, 5, 6, 7, 18, 17, 16);
// NOTE taht the pins have bee rearranged
U8GLIB_ST7920_128X64_4X u8g(8, 9, 10, 11, 4, 5, 6, 7, 1, 2, 3);

byte Sample[100];
//int Sample[100];
int Input=0;
int OldInput=0;
long StartSample=0;
long EndSample=0;
long SampleTime=0;
int MaxSample=0;
int MinSample=0;
int SampleSize=0;
int dTime=0;
int mode=0;

void u8g_prepare(void) {
  u8g.setFont(u8g_font_6x10);
  u8g.setFontRefHeightExtendedText();
  u8g.setDefaultForegroundColor();
  u8g.setFontPosTop();
}

void DrawMarkers(void) {
```

```
  u8g.drawFrame (0,0,128,64);
  u8g.drawFrame (100,0,128,64);
  u8g.drawPixel (25,16);
  u8g.drawPixel (50,16);
  u8g.drawPixel (75,16);
  u8g.drawPixel (25,32);
  u8g.drawPixel (50,32);
  u8g.drawPixel (75,32);
  u8g.drawPixel (25,48);
  u8g.drawPixel (50,48);
  u8g.drawPixel (75,48);
}

void get_mode(void) {
   if (digitalRead(13) == 0) mode++;
   if (digitalRead(12) == 0) mode--;
   if (mode > 9) mode = 0;
   if (mode < 0) mode = 9;
// Select delay times for loop modes
   if (mode == 0) dTime=0;
   if (mode == 1) dTime=0;
   if (mode == 2) dTime=1;
   if (mode == 3) dTime=5;
   if (mode == 4) dTime=10;
   if (mode == 5) dTime=50;
   if (mode == 6) dTime=100;
   if (mode == 7) dTime=500;
   if (mode == 8) dTime=1000;
   if (mode == 9) dTime=5000;
}
void sample_data(void){
// wait for a trigger of a positive going input
  while (digitalRead(A0)==0) {  }
// collect the analog data into an array
// mode 0 will use verbose method
if (mode == 0) {
  StartSample = micros();
   Sample[0]=PINC;
   Sample[1]=PINC;    Sample[2]=PINC;    Sample[3]=PINC;
   Sample[4]=PINC;    Sample[5]=PINC;    Sample[6]=PINC;
   Sample[7]=PINC;    Sample[8]=PINC;    Sample[9]=PINC;
```

```
    Sample[10]=PINC;    Sample[11]=PINC;    Sample[12]=PINC;
    Sample[13]=PINC;    Sample[14]=PINC;    Sample[15]=PINC;
    Sample[16]=PINC;    Sample[17]=PINC;    Sample[18]=PINC;
    Sample[19]=PINC;    Sample[20]=PINC;    Sample[21]=PINC;
    Sample[22]=PINC;    Sample[23]=PINC;    Sample[24]=PINC;
    Sample[25]=PINC;    Sample[26]=PINC;    Sample[27]=PINC;
    Sample[28]=PINC;    Sample[29]=PINC;    Sample[30]=PINC;
    Sample[31]=PINC;    Sample[32]=PINC;    Sample[33]=PINC;
    Sample[34]=PINC;    Sample[35]=PINC;    Sample[36]=PINC;
    Sample[37]=PINC;    Sample[38]=PINC;    Sample[39]=PINC;
    Sample[40]=PINC;    Sample[41]=PINC;    Sample[42]=PINC;
    Sample[43]=PINC;    Sample[44]=PINC;    Sample[45]=PINC;
    Sample[46]=PINC;    Sample[47]=PINC;    Sample[48]=PINC;
    Sample[49]=PINC;    Sample[50]=PINC;    Sample[51]=PINC;
    Sample[52]=PINC;    Sample[53]=PINC;    Sample[54]=PINC;
    Sample[55]=PINC;    Sample[56]=PINC;    Sample[57]=PINC;
    Sample[58]=PINC;    Sample[59]=PINC;    Sample[60]=PINC;
    Sample[61]=PINC;    Sample[62]=PINC;    Sample[63]=PINC;
    Sample[64]=PINC;    Sample[65]=PINC;    Sample[66]=PINC;
    Sample[67]=PINC;    Sample[68]=PINC;    Sample[69]=PINC;
    Sample[70]=PINC;    Sample[71]=PINC;    Sample[72]=PINC;
    Sample[73]=PINC;    Sample[74]=PINC;    Sample[75]=PINC;
    Sample[76]=PINC;    Sample[77]=PINC;    Sample[78]=PINC;
    Sample[79]=PINC;    Sample[80]=PINC;    Sample[81]=PINC;
    Sample[82]=PINC;    Sample[83]=PINC;    Sample[84]=PINC;
    Sample[85]=PINC;    Sample[86]=PINC;    Sample[87]=PINC;
    Sample[88]=PINC;    Sample[89]=PINC;    Sample[90]=PINC;
    Sample[91]=PINC;    Sample[92]=PINC;    Sample[93]=PINC;
    Sample[94]=PINC;    Sample[95]=PINC;    Sample[96]=PINC;
    Sample[97]=PINC;    Sample[98]=PINC;    Sample[99]=PINC;
//    Sample[100]=PINC;
   EndSample = micros();
 }
// mode 1 will use loop with no delay
if (mode ==1) {
  StartSample = micros();
   for(int xpos=0; xpos<100; xpos++) {
    Sample[xpos]=PINC;
//    delayMicroseconds(dTime);
   }
   EndSample = micros();
```

```
}
// mode 2 or more will use loop with delay
if (mode >= 2) {
  StartSample = micros();
  for(int xpos=0; xpos<100; xpos++) {
    Sample[xpos]=PINC;
    delayMicroseconds(dTime);
  }
  EndSample = micros();
} }
void draw(void) {
  char buf[12];
  u8g_prepare();
  DrawMarkers();
// display the collected analog data from array
  for(int xpos=1; xpos<99; xpos++) {
// For Oscope more use this line
    u8g.drawLine (xpos, Sample[xpos], xpos+1, Sample[xpos+1]);
// For logic analizer use the next 6 lines instead
//    u8g.drawLine (xpos, ((Sample[xpos]&B00000001)*4)+4, xpos, ((Sample[xpos+1]&B00000001)*4)+4);
//    u8g.drawLine (xpos, ((Sample[xpos]&B00000010)*2)+14, xpos, ((Sample[xpos+1]&B00000010)*2)+14);
//    u8g.drawLine (xpos, ((Sample[xpos]&B00000100)*1)+24, xpos, ((Sample[xpos+1]&B00000100)*1)+24);
//    u8g.drawLine (xpos, ((Sample[xpos]&B00001000)/2)+34, xpos, ((Sample[xpos+1]&B00001000)/2)+34);
//    u8g.drawLine (xpos, ((Sample[xpos]&B00010000)/4)+44, xpos, ((Sample[xpos+1]&B00010000)/4)+44);
//    u8g.drawLine (xpos, ((Sample[xpos]&B00100000)/8)+54, xpos, ((Sample[xpos+1]&B00100000)/8)+54);
  }
  SampleTime=EndSample-StartSample;
  if (SampleTime < 9999) u8g.drawStr(102, 2, "uS");
  if (SampleTime > 9999) {
    SampleTime=SampleTime/1000;
    u8g.drawStr(102, 2, "mS");
  }
  u8g.drawStr(102, 12, itoa(SampleTime, buf, 10));
  u8g.drawStr(102, 22, "Mode");
  u8g.drawStr(102, 32, itoa(mode, buf, 10));
```

```
// Determine sample voltage peak to peak
  MaxSample = Sample[10];
  MinSample = Sample[10];
  for(int xpos=0; xpos<100; xpos++) {
//    OldSample[xpos] = Sample[xpos];
    if (Sample[xpos] > MaxSample) MaxSample=Sample[xpos];
    if (Sample[xpos] < MinSample) MinSample=Sample[xpos];
    }
  // Range of 0 to 64 * 78 = 4992 mV
  SampleSize=(MaxSample-MinSample)*78;
  u8g.drawStr(102, 42, "mV");
  u8g.drawStr(102, 52, itoa(SampleSize, buf, 10));
}
void setup(void) {
  // set up input pins
  pinMode(12, INPUT);
  digitalWrite(12, HIGH);
  pinMode(13, INPUT);
  digitalWrite(13, HIGH);
  // assign default color value
  if ( u8g.getMode() == U8G_MODE_R3G3B2 )
    u8g.setColorIndex(255);     // RGB=white
  else if ( u8g.getMode() == U8G_MODE_GRAY2BIT )
    u8g.setColorIndex(3);       // max intensity
  else if ( u8g.getMode() == U8G_MODE_BW )
    u8g.setColorIndex(1);       // pixel on, black
}
void loop(void) {
// Set up the mode
  get_mode();
// collect the data
  sample_data();
// show collected data
  u8g.firstPage();
  do { draw(); }
  while( u8g.nextPage() );
// rebuild the picture after some delay
  delay(500);
}
```

Chapter 9

Medium Resolution Graphic LCD

1.8TFT SPI 128 by 160

Someone pointed out that this LCD screen was left out of earlier editions of my LCD book, so I thought I would add it later. There appears to be two versions of the 1.8 inch TFT LCD screen. One version has 10 pins and the other version has 16 pins. The pin definitions are written on the bottom of the circuit board so it is just a matter of writing them down before you flip it over and wire it up.

Here is a picture of the back of the 1.8 inch LCD. As you can see the pins are clearly marked.

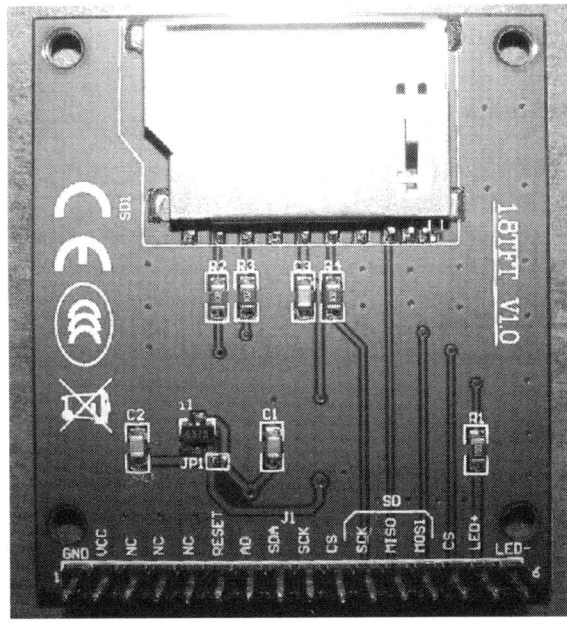

The sketches for this LCD use the new built in TFT drivers found in version 1.0.5 of the Arduino driver. They will not work without this TFT driver being properly installed. This built in TFT driver is the same one as the adafruit ST7735 driver, it is just renamed as "TFT".

Here is a chart showing the wiring from the Arduino Uno to the LCD screen. I ran jumpers to connect the two ground and 5V pins together on the breadboard.

```
Arduino Uno          1.8 SPI TFT
---------------      ---------------
GND                  Pin 01 (GND)
5V (VCC)             Pin 02 (VCC)
Not used             Pin 03-05
D8                   Pin 06 (RESET)
D9                   Pin 07 (A0)
D11 (MOSI)           Pin 08 (SDA)
D13 (SCK)            Pin 09 (SCK)
D10 (SS)             Pin 10 (CS)
SD Card              Pins 11-14
5V (VCC)             Pin 15 (LED+)
GND                  Pin 16 (LED-)
```

Note that pin one of the LCD is on the right as you look at the top of the LCD screen.

This is the schematic diagram of how to wire it up. I suspect that some of the wires shown are not actually needed as only three Arduino pins are actually used in the sketch.

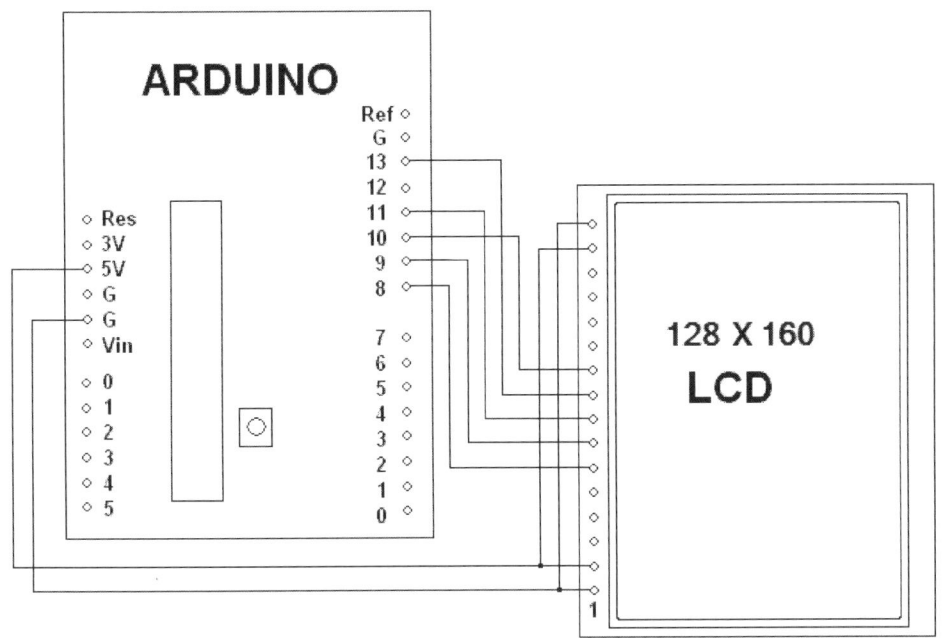

Up next is a picture of the LCD wired up and displaying an analog meter simulation.

This sketch is an analog meter demonstration.

/**********************************
TFT Analog Meter
Reads the value of analog sensor on A0,
and shows the value on the screen.

Created 15 January 2014 by Bob Davis
**************************************/

```c
#include <TFT.h>  // Arduino LCD library
#include <SPI.h>

 // pin definition for the Uno
#define rst  8
#define dc   9
#define cs   10

TFT TFTscreen = TFT(cs, dc, rst);

// set up variables
int xPos = 0;
int sensor = 0;
int drawLine = 0;
char buf[12];

void setup(){
  // initialize the display
  TFTscreen.begin();
  // clear the screen
  TFTscreen.background(250, 250, 250);
  // Set the font size
  TFTscreen.setTextSize(2);
}

void loop(){
  // read the sensor
  sensor = analogRead(A0);
  // Map the sensor off sides for wider movement
  drawLine = map(sensor, 0, 1023, -60, 220);
  // select the color = B,G,R
  TFTscreen.stroke(0, 0, 250);
  // draw the line (xPos1, yPos1, xPos2, yPos2);
  TFTscreen.line(80, 128, drawLine, 0);
  // Set font color to blue
  TFTscreen.stroke(255, 0, 0);
  // Write the text value of the sensor
  TFTscreen.text( itoa(sensor/2, buf, 10), 10, 110);
```

    delay(200);
    // erase the screen and start again
    TFTscreen.background(255, 255, 255);
}

Up next is a picture of this LCD screen while running the oscilloscope demonstration sketch. This is a fairly simple oscilloscope. For an analog input circuit see the other oscilloscope articles in this book.

This sketch produces a simple oscilloscope that is good to about 1KHz. The trace is in red and the text listing the peak to peak voltage is in green.

/***********************************
 TFT Oscope
 Reads the value of analog input on A0,
 and shows the value on the screen.
 Created 15 January 2014 by Bob Davis
 ***************************************/

#include <TFT.h> // Arduino LCD library
#include <SPI.h>

 // pin definition for the Uno
#define rst 8
#define dc   9

```
#define cs   10

TFT TFTscreen = TFT(cs, dc, rst);

// set up variables
int xPos = 0;
int value = 0;
int maxvalue=100;
int minvalue=100;
int sensor[160];
char buf[12];

void setup(){
  // initialize the display
  TFTscreen.begin();
  // clear the screen
  TFTscreen.background(0, 0, 0);
  // Set the font size
  TFTscreen.setTextSize(2);
}

void loop(){
  // quickly collect the data
  for (int xpos = 0; xpos <160; xpos++){
  sensor[xpos] = analogRead(A0);
  }
  // determine the peak to peak voltage
  maxvalue=sensor[1];
  minvalue=sensor[1];
  for (int xpos = 0; xpos <160; xpos++){
  if (sensor[xpos] > maxvalue) maxvalue=sensor[xpos];
  if (sensor[xpos] < minvalue) minvalue=sensor[xpos];
  }
  value=maxvalue-minvalue;
  // erase the screen to start again
  TFTscreen.background(0, 0, 0);
  // display the collected data
  for (int xpos = 0; xpos <159; xpos++){
  // select the color = B,G,R
  TFTscreen.stroke(50, 50, 255);
  // draw the line (xPos1, yPos1, xPos2, yPos2);
```

```
TFTscreen.line(xpos, sensor[xpos]/8, xpos+1, sensor[xpos+1]/8);
// Set font color to green
TFTscreen.stroke(0, 255, 0);
// Write the text value of the sensor
if (xpos==0) TFTscreen.text( itoa(value/2, buf, 10), 10, 110);
if (xpos==0) TFTscreen.text( "mv", 50, 110);
 }
}
```

Chapter 10

High Resolution Graphic LCD

320 by 240 TFT240_262K

When it comes to higher resolution LCD's there are many of them to choose from. They vary in size from 1.8 inches to 2.4 inches to 2.8 inches to 3.2 inches. Many even larger sizes are also available. How do you choose a LCD to work with? First, look at the number of pins on the LCD's connector. The larger ones have two rows of what looks like about 20 pins each because they are 16 bit devices. This will not plug into most breadboards, and there are too many pins for an Arduino UNO. However if you have an Arduino Mega or what is called a "Funduino", then you have a connector that perfectly matches the one that is found on the larger LCD's.

There are also a number of LCD's available on eBay that only have a flexible flat ribbon cable going to them. If you do not have a jack that matches that flat ribbon cable, or an adapter circuit board, you cannot use this type of LCD either.

The one I picked is labeled HY-TFT240_262K, it has a 18 pin connector for the LCD data. It also has connectors for the touch screen and for the on board memory module. We are just concerned with the 18 pin connector for now. Some of these LCD's have a jumper for 3.3 or five volt operation. Make sure that any jumpers are set properly before powering on the LCD.

For this project we will be using the "UTFT library". That stands for "Universal Thin Film Transistor" library. These LCD displays are referred to as "TFT" or "Thin Film Transistor" LCD displays. They feature three color operation as well as higher resolutions. They were designed for use in cell phones and MP4 players. The UTFT library is downloaded as "UTFT.rar". Because UTFT.rar is a "rar" file, and not a "zip" file, you might have to also download a "rar" compatible decompression program like "7-Zip" to decompress it. Once again, you place the decompressed files in your "Arduino\Libraries" directory. It should look like the picture below once it is properly installed.

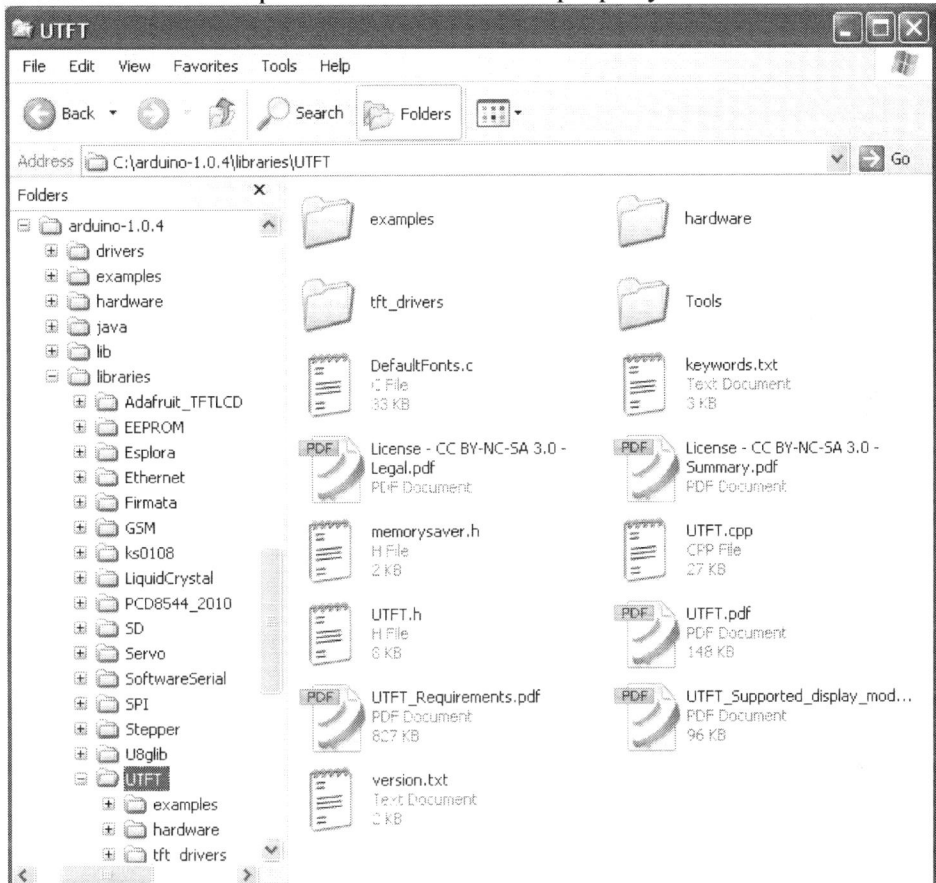

Here is a list of the UTFT Commands

UTFT (Model, RS, WR, CS, RST, ALE);
InitLCD(Orientation) ;
getDisplayXSize();
getDisplayYSize();
lcdOff();
lcdOn();
setContrast(c);
clrScr();
fillScr(r, g, b); fillScr(color);

setColor(r, g, b); setColor(color);
getColor();
setBackColor(r, g, b); setBackColor(color);
getBackColor();
drawPixel(x, y);
drawLine(x1, y1, x2, y2);
drawRect(x1, y1, x2, y2);
drawRoundRect(x1, y1, x2, y2);
fillRect(x1, y1, x2, y2);
fillRoundRect(x1, y1, x2, y2);
drawCircle(x, y, radius);
fillCircle(x, y, radius);
print(st, x, y, deg);
printNumI(num, x, y, length, filler);
printNumF(num, dec, x, y, divider, length, filler);
setFont(fontname);
getFont();
getFontXsize();
getFontYsize();

Here is a list of some of the many chip sets that are supported by the UTFT library. Note that many of the supported chips are available in 8 bit, 16 bit and serial versions.

The Arduino UNO can be used with 8 bit and serial versions.

| | | Supported mode | | |
Controller	Model for UTFT()	8bit	16bit	Serial
HX8340-B(N)	HX8340B_S			✓
HX8340-B(T)	HX8340B_8	✓		
HX8347-A	HX8347A		✓	
HX8352-A	HX8352A		✓	
ILI9320	ILI9320_8	✓		
	ILI9320_16		✓	
ILI9325C	ILI9325C	✓		
ILI9325D	ILI9325D_8	✓		
	ILI9325D_16		✓	
ILI9327	ILI9327		✓	
ILI9481	ILI9481		✓	
PCF8833	PCF8833			✓
S1D19122	S1D19122		✓	
S6D1121	S6D1121_8	✓		
	S6D1121_16		✓	
SSD1289	SSD1289		✓	
	SSD1289_8	✓		
	SSD1289LATCHED[6]		LATCHED	
SSD1963	SSD1963_480		✓	
	SSD1963_800		✓	
	SSD1963_800ALT[7]		✓	
ST7735	ST7735			✓

According to the UTFT library manual in PDF format, this is how to connect the supported TFT LCD's to the Arduino processors. Note that DB0 to DB7 become DB8 to DB15 for 8 bit interfaces.

This list also shows how to connect a 16 bit device to the Arduino UNO but that uses every last available pin. The TFT pin listing does not match our TFT display. The pinout listed is for 16 bit displays.

Signal	TFT pin	Arduino	
		2009/Uno/Leonardo	Mega/Due[2]
DB0[5]	21	D8	D37
DB1[5]	22	D9	D36
DB2[5]	23	D10	D35
DB3[5]	24	D11	D34
DB4[5]	25	D12	D33
DB5[5]	26	D13	D32
DB6[5]	27	A0 (D14)	D31
DB7[5]	28	A1 (D15)	D30
DB8	7	D0	D22
DB9	8	D1	D23
DB10	9	D2	D24
DB11	10	D3	D25
DB12	11	D4	D26
DB13	12	D5	D27
DB14	13	D6	D28
DB15	14	D7	D29
RS	4	Any free pin	
WR	5	Any free pin	
RD	6	Must be pulled high (3.3v)	
CS	15	Any free pin	
REST	17	Any free pin	

Here is another pinout for connecting the 18 pin LCD to the Arduino UNO.

LCD Pin	Name	Arduino
1	Gnd	Gnd
2	+5V	5V
3	NC	
4	RS	A5
5	RW	A4
6	RD	3.3V
7	DB0	D0 (May cause issues)
8	DB1	D1
9	DB2	D2
10	DB3	D3
11	DB4	D4
12	DB5	D5
13	DB6	D6

14	DB7	D7
15	CS	A3
16		
17	Rst	A2
18		

Here is a schematic diagram of how to connect it all up.

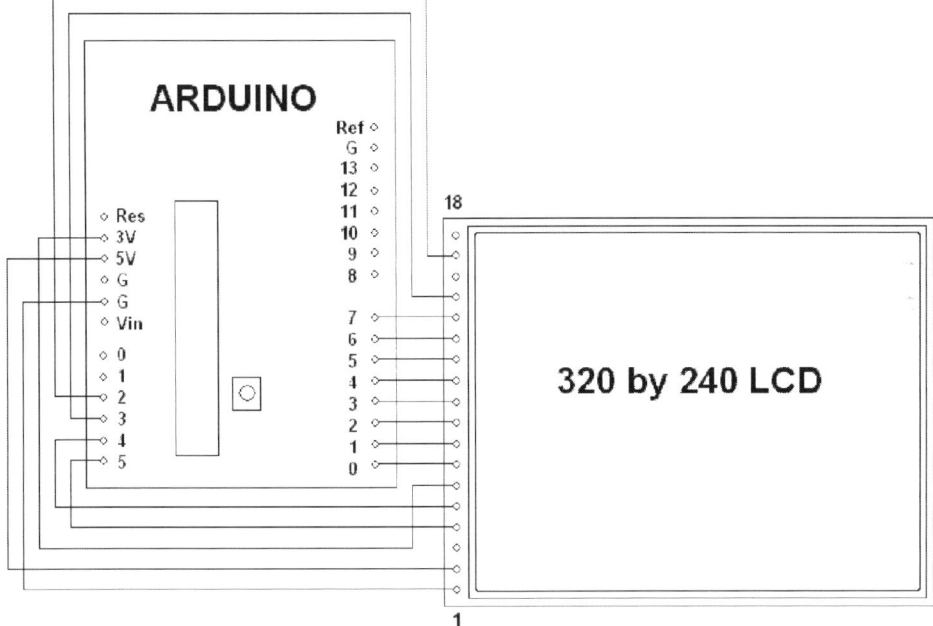

Up next is a picture of the analog meter simulation sketch running. It features a meter history in red, and the live meter in white. There is also a green text display of the input voltage in millivolts.

If you connect the sound output from your computer to the analog input you can watch the meter dance with the music once again.

Here is the sketch code for the three color analog meter simulation.

```
//************************************
// Three color analog meter simulation
// By Bob Davis
// UTFT_(C)2012 Henning Karlsen
// web: http://www.henningkarlsen.com/electronics
//

#include <UTFT.h>
// Declare which fonts we will be using
extern uint8_t SmallFont[];
extern uint8_t BigFont[];
extern uint8_t SevenSegNumFont[];

UTFT myGLCD(ILI9325C,19,18,17,16);
int Mv;
int OldMv;

void setup() {
  myGLCD.InitLCD();
  myGLCD.clrScr();
}
```

```
void loop() {
  // set color(Red, Green, Blue) range = 0 to 255
  myGLCD.setBackColor(0, 0, 0);
  myGLCD.setFont(BigFont);
  char buf[12];
  while(1) {
    // display millivolts in green text
    Mv=analogRead(A0)/2.05;
    myGLCD.setColor(0, 255, 0);
    myGLCD.print(itoa(Mv, buf, 10), 16, 200);
    // Turn old line red
    if (OldMv != Mv) {
      myGLCD.setColor(255, 0, 0);
      myGLCD.drawLine(160, 240, OldMv, 0);
    }
    // Draw new white line
    myGLCD.setColor(255, 255, 255);
    myGLCD.drawLine(160, 240, Mv, 0);
    OldMv=Mv;
  }; // do nothing
}
```

Above is a picture of the oscilloscope program running. It had green division markers, blue text showing the millivolts peak to peak, and blue text showing the sample time. A switch connected to A1 can select fast sampling when ground, or slow sampling when 5V. A software trigger loop makes the waveform fairly stable. For the analog input circuit schematic, see the oscilloscope input schematic for the 128 by 64 LCD.

Here is the sketch for the color Arduino oscilloscope

```
//*************************************
// Three color Fast analog Oscilloscope
// By Bob Davis
// UTFT_(C)2012 Henning Karlsen
// web: http://www.henningkarlsen.com/electronics
//

#include <UTFT.h>
#ifdef cbi
#define cbi(sfr, bit) (_SFR_BYTE(sfr) &= ~_BV(bit))
#endif
#ifdef sbi
#define sbi(sfr, bit) (_SFR_BYTE(sfr) |= _BV(bit))
#endif

// Declare which fonts we will be using
extern uint8_t SmallFont[];
extern uint8_t BigFont[];
extern uint8_t SevenSegNumFont[];

UTFT myGLCD(ILI9325C,19,18,17,16);
int Input=0;
int OldInput=0;
int MaxSample=0;
int MinSample=0;
int Sample[320];
int OldSample[320];
int StartTime=0;
int EndTime=0;
int rate=1;

void DrawMarkers(){
  myGLCD.setColor(0, 200, 0);
  myGLCD.drawLine(0, 0, 0, 240);
  myGLCD.drawLine(54, 0, 54, 240);
  myGLCD.drawLine(107, 0, 107, 240);
```

```
  myGLCD.drawLine(160, 0, 160, 240);
  myGLCD.drawLine(213, 0, 213, 240);
  myGLCD.drawLine(266, 0, 266, 240);
  myGLCD.drawLine(319, 0, 319, 240);
  myGLCD.drawLine(0, 0, 319, 0);
  myGLCD.drawLine(0, 50, 319, 50);
  myGLCD.drawLine(0, 100, 319, 100);
  myGLCD.drawLine(0, 150, 319, 150);
  myGLCD.drawLine(0, 200, 319, 200);
  myGLCD.drawLine(0, 239, 319, 239);
}

void setup() {
  myGLCD.InitLCD();
  myGLCD.clrScr();
  myGLCD.setBackColor(0, 0, 0);
  myGLCD.setFont(BigFont);
}

void loop() {
  // set color(Red, Green, Blue) range = 0 to 255
  char buf[12];
  while(1) {
  // Set sample speed according to switch on A1
  if (analogRead(A1) < 500){
    cbi(ADCSRA, ADPS2);
    rate=0;
  }
  else {
    sbi(ADCSRA, ADPS2);
    rate=1;
  }
  DrawMarkers();
  // wait for a trigger of a positive going input
  OldInput=analogRead(A0);
  Input=analogRead(A0);
  while (Input < OldInput){
    Input=analogRead(A0);
  }
  // collect the analog data into an array
  // do not do division here, it makes it slower!
```

```
  StartTime = micros();
  for(int xpos=0; xpos<319; xpos++) {
    Sample[xpos]=analogRead(A0);
  }
  EndTime = micros();
  // display the collected analog data from array
  // Sample/4.1 because 1024 becomes 250 = 5 volts
  for(int xpos=0; xpos<319; xpos++) {
    myGLCD.setColor(0, 0, 0);
    myGLCD.drawLine (xpos, OldSample[xpos]/4.1, xpos+1, OldSample[xpos+1]/4.1);
    myGLCD.setColor(255, 255, 255);
    myGLCD.drawLine (xpos, Sample[xpos]/4.1, xpos+1, Sample[xpos+1]/4.1);
  }
// Determine sample voltage peak to peak
  MaxSample = Sample[100];
  MinSample = Sample[100];
  for(int xpos=0; xpos<319; xpos++) {
    OldSample[xpos] = Sample[xpos];
    if (Sample[xpos] > MaxSample) MaxSample=Sample[xpos];
    if (Sample[xpos] < MinSample) MinSample=Sample[xpos];
  }
// Display sample voltage and time
  myGLCD.setColor(0, 0, 255);
  int SampleSize=MaxSample-MinSample;
  myGLCD.print("MV=", 1, 220);
  myGLCD.print(itoa(SampleSize, buf, 10), 44, 220);
  int SampleTime=EndTime-StartTime;
  myGLCD.print("uS=     ", 161, 220);
  // Adjust time according to sample speed
  if (rate==1) {
    SampleTime=(EndTime/1000-StartTime/1000);
    myGLCD.print("mS=     ", 161, 220);
  }
  myGLCD.print(itoa(SampleTime, buf, 10), 210, 220);
  }
}
```

Here is a close up picture of the two channel logic analyzer screen. The samples are of 40 Kc and 400 Kc Pulses. The schematic is the same as the oscilloscope. The digital inputs are A0 and A1. The sample rate is 2 million samples per second.

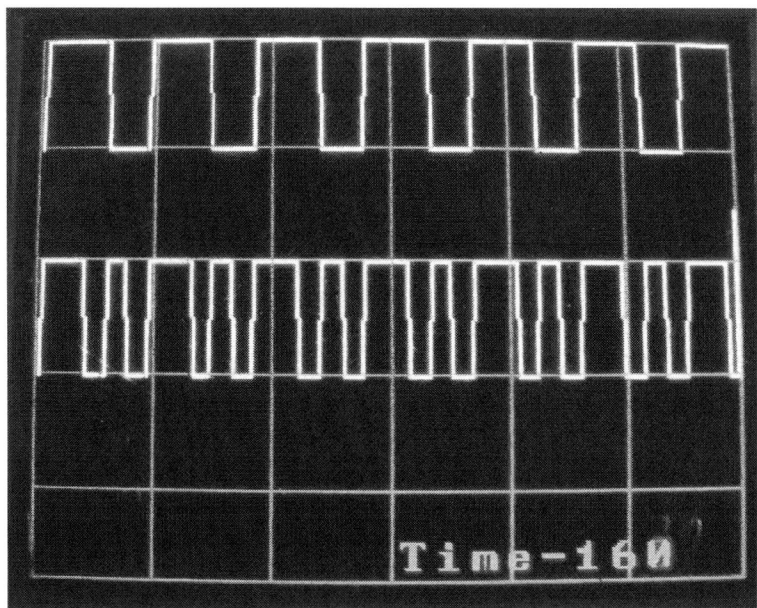

Here is a sketch for a fast two channel logic analyzer. It uses the parallel input command, "PIN" for the analog port "C"

```
//*****************************************
// Three color FAST Logic Analizer
// By Bob Davis
// UTFT_(C)2012 Henning Karlsen
// web: http://www.henningkarlsen.com/electronics
//

#include <UTFT.h>
// Declare which fonts we will be using
extern uint8_t SmallFont[];
extern uint8_t BigFont[];
extern uint8_t SevenSegNumFont[];

UTFT myGLCD(ILI9325C,19,18,17,16);
int Input=0;
int OldInput=0;
```

```
byte Sample[320];
byte Sample0[320];
byte OldSample0[320];
byte Sample1[320];
byte OldSample1[320];
int StartSample=0;
int EndSample=0;

void DrawMarkers(){
  myGLCD.setColor(0, 200, 0);
  myGLCD.drawLine(0, 0, 0, 240);
  myGLCD.drawLine(54, 0, 54, 240);
  myGLCD.drawLine(107, 0, 107, 240);
  myGLCD.drawLine(160, 0, 160, 240);
  myGLCD.drawLine(213, 0, 213, 240);
  myGLCD.drawLine(266, 0, 266, 240);
  myGLCD.drawLine(319, 0, 319, 240);
  myGLCD.drawLine(0, 0, 319, 0);
  myGLCD.drawLine(0, 50, 319, 50);
  myGLCD.drawLine(0, 100, 319, 100);
  myGLCD.drawLine(0, 150, 319, 150);
  myGLCD.drawLine(0, 200, 319, 200);
  myGLCD.drawLine(0, 239, 319, 239);
}

void setup() {
  myGLCD.InitLCD();
  myGLCD.clrScr();
  pinMode(14, INPUT);
  pinMode(15, INPUT);
}
void loop() {
  // set color(Red, Green, Blue) range = 0 to 255
  myGLCD.setBackColor(0, 0, 0);
  myGLCD.setFont(BigFont);
  char buf[12];
  while(1) {
  DrawMarkers();
  // wait for trigger of a positive input
  Input=digitalRead(A0);
  while (Input == 0){
```

```
    Input=digitalRead(A0);
  }
  // collect the data into arrays
  // Read analog port as a parallel port PINC
  StartSample = micros();
  for(int xpos=0; xpos<319; xpos++) {
    Sample[xpos]=PINC;
  }
  EndSample = micros();
  // make data into displayable information
  // convert binary data by masking out bits
     for(int xpos=0; xpos<319; xpos++) {
     Sample0[xpos] = (((Sample[xpos] & B00000001)*50)+1);
     Sample1[xpos] = (((Sample[xpos] & B00000010)*25)+101);
     }
  // display the collected data from array
   for(int xpos=0; xpos<319; xpos++) {
     myGLCD.setColor(0, 0, 0);
     myGLCD.drawLine (xpos, OldSample0[xpos], xpos+1,
OldSample0[xpos+1]);
     myGLCD.drawLine (xpos, OldSample1[xpos], xpos+1,
OldSample1[xpos+1]);
     myGLCD.setColor(255, 255, 255);
     myGLCD.drawLine (xpos, Sample0[xpos], xpos+1, Sample0[xpos+1]);
     myGLCD.drawLine (xpos, Sample1[xpos], xpos+1, Sample1[xpos+1]);
     }
  // store samples to oldsamples
   for(int xpos=0; xpos<319; xpos++) {
     OldSample0[xpos] = Sample0[xpos];
     OldSample1[xpos] = Sample1[xpos];
     }
  // display the sample time
   myGLCD.setColor(0, 0, 255);
   int SampleTime=StartSample-EndSample;
   myGLCD.print("Time=", 161, 220);
   myGLCD.print(itoa(SampleTime, buf, 10), 224, 220);
     }
}
```

Of course, I will also show you how to make a six channel logic analyzer using this LCD display. You can use a loop to gather the data via "PINC" at about 1.33 million samples per second. You could also say "Sample[1]=PINC, Sample[2]=PINC, Sample[3]=PINC" etc, and then you can reach a speed of about 3 million samples per second. You can also remove the Boolean statements under "// draw new data" and instead say "myGLCD.drawLine (xpos, Sample[xpos], xpos+1, Sample[xpos+1])" and you will then have an oscilloscope when a fast "flash" type of external analog to digital converter is connected.

Here is the modified schematic with the changes needed to free up the analog input pins.

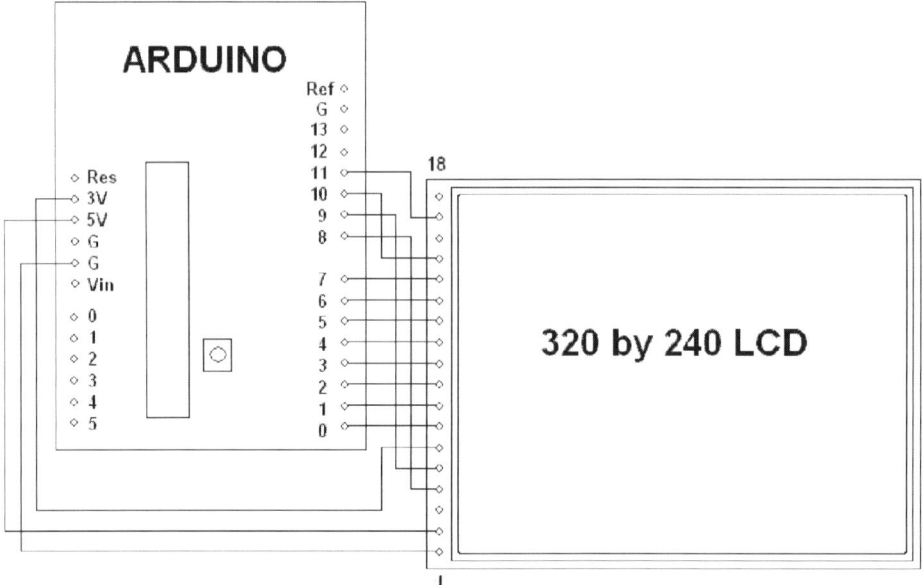

Up next is a picture of the six channel logic analyzer. It is showing the six outputs of an AD775 fast analog to digital converter with a 20 KC sine wave on its input.

Here is the code for a sketch to make the logic analyzer work.

```
//****************************************
// Three color 6 channel Logic Analyzer
// By Bob Davis
// UTFT_(C)2012 Henning Karlsen
// web: http://www.henningkarlsen.com/electronics
//

#include <UTFT.h>

// Declare which fonts we will be using
extern uint8_t SmallFont[];
extern uint8_t BigFont[];
extern uint8_t SevenSegNumFont[];

// Note that control pins are now assigned to pins 8-11
UTFT myGLCD(ILI9325C,8,9,10,11);
int Input=0;
int Sample[320];
int StartSample=0;
int EndSample=0;

void DrawMarkers(){
  myGLCD.setColor(0, 200, 0);
  myGLCD.drawLine(0, 0, 0, 240);
```

```
    myGLCD.drawLine(54, 0, 54, 240);
    myGLCD.drawLine(107, 0, 107, 240);
    myGLCD.drawLine(160, 0, 160, 240);
    myGLCD.drawLine(213, 0, 213, 240);
    myGLCD.drawLine(266, 0, 266, 240);
    myGLCD.drawLine(319, 0, 319, 240);
    myGLCD.drawLine(0, 0, 319, 0);
    myGLCD.drawLine(0, 50, 319, 50);
    myGLCD.drawLine(0, 100, 319, 100);
    myGLCD.drawLine(0, 150, 319, 150);
    myGLCD.drawLine(0, 200, 319, 200);
    myGLCD.drawLine(0, 239, 319, 239);
}
void setup() {
  myGLCD.InitLCD();
  myGLCD.clrScr();
  pinMode(14, INPUT);
  pinMode(15, INPUT);
  pinMode(16, INPUT);
  pinMode(17, INPUT);
  pinMode(18, INPUT);
  pinMode(19, INPUT);
}
void loop() {
  // set color(Red, Green, Blue) range = 0 to 255
  myGLCD.setBackColor(0, 0, 0);
  myGLCD.setFont(BigFont);
  char buf[12];
  while(1) {
  DrawMarkers();
  // wait for trigger of a positive input
  while (Input == 0){
    Input=digitalRead(A0);
    }
// collect the analog data into an array
// Read analog port as a parallel port PINC
  StartSample = micros();
  for(int xpos=0; xpos<319; xpos++) {
    Sample[xpos]=PINC;
  }
  EndSample = micros();
```

```
// display the collected analog data from array
  for(int xpos=0; xpos<319; xpos++) {
    // Erase old stuff
    myGLCD.setColor(0, 0, 0);
    myGLCD.drawLine (xpos+1, 1, xpos+1, 220);
    // Draw new data
    myGLCD.setColor(255, 255, 255);
    myGLCD.drawLine (xpos, ((Sample[xpos]&B00000001)*16)+2,
xpos+1, ((Sample[xpos+1]&B00000001)*16)+2);
    myGLCD.drawLine (xpos, ((Sample[xpos]&B00000010)*8)+42,
xpos+1, ((Sample[xpos+1]&B00000010)*8)+42);
    myGLCD.drawLine (xpos, ((Sample[xpos]&B00000100)*4)+82,
xpos+1, ((Sample[xpos+1]&B00000100)*4)+82);
    myGLCD.drawLine (xpos, ((Sample[xpos]&B00001000)*2)+122,
xpos+1, ((Sample[xpos+1]&B00001000)*2)+122);
    myGLCD.drawLine (xpos, ((Sample[xpos]&B00010000)/1)+162,
xpos+1, ((Sample[xpos+1]&B00010000)/1)+162);
    myGLCD.drawLine (xpos, ((Sample[xpos]&B00100000)/2)+202,
xpos+1, ((Sample[xpos+1]&B00100000)/2)+202);
  }
  // display the sample time
  myGLCD.setColor(0, 0, 255);
  int SampleTime=EndSample-StartSample;
  myGLCD.print("MicroSeconds=", 10, 220);
  myGLCD.print(itoa(SampleTime, buf, 10), 224, 220);
    }
}
```

We can even make a faster oscilloscope if we add an external analog to digital converter such as the CA3306. It is a "flash" converter meaning that it has 64 comparators that instantly converts the analog input to a digital output. Because of that, it can do over 15 million conversions per second.

Here is the schematic of the CA3306 analog to digital converter. The schematic is adapted from the PDF specifications file for the CA3306. The clock input can come from any four MC to 15 MC clock oscillator.

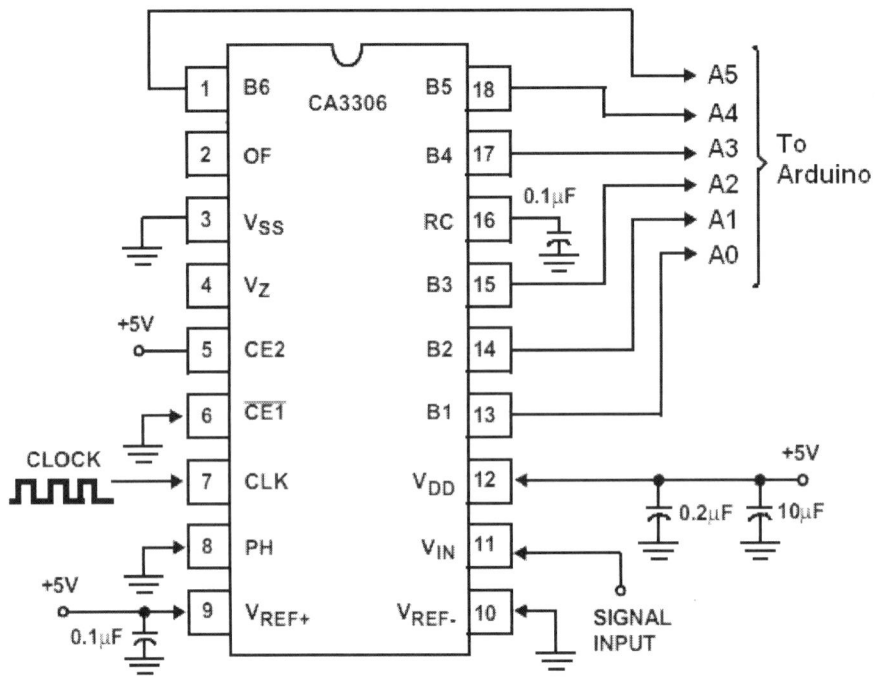

Here is a picture of the CA3306 analog to digital converter with an oscillator and filter capacitors.

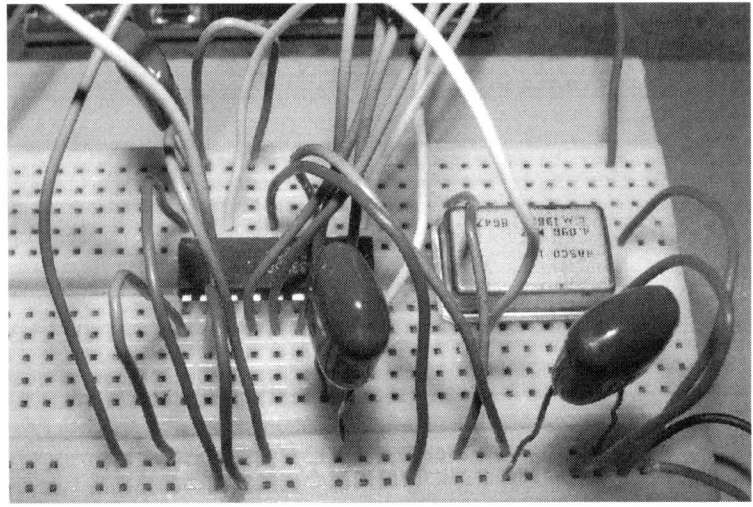

A better analog input section can also be added to provide either input attenuation of 1/10 for a 50 volt signal or a gain of up to 5X with just one IC, either a LF353 or a TL082. The drawback of the improved input section is that you will need a positive and negative 9 to 12 volt power source. Two 9 volt batteries will work. This improved input circuit will also present a smaller load, of one million ohms, to the circuit that is under test.

Coming up next is the schematic diagram of the improved analog input section. It is very basic, but does provide some selection of the input attenuation or of the gain, as well as diode over voltage protection, for the input and for the analog to digital converter.

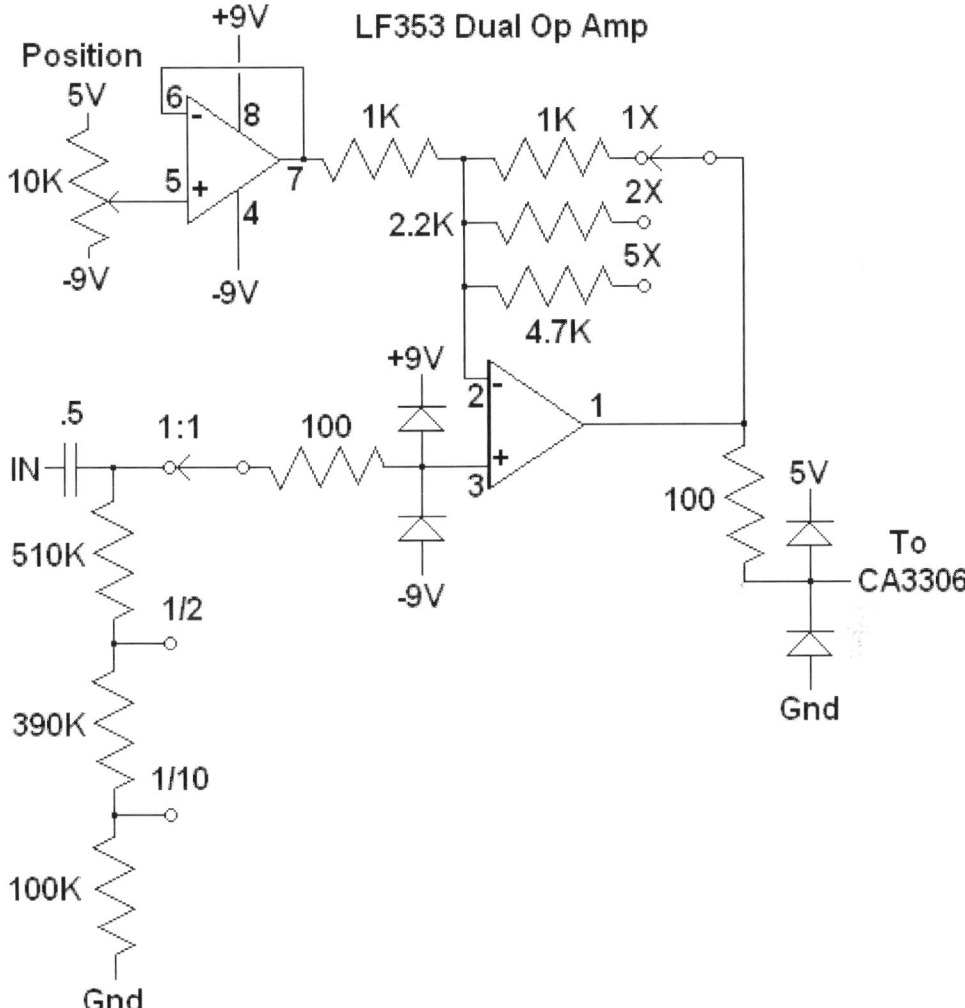

Here is the sketch code to make it all work. Note that two momentary contact push buttons are added. One goes to D12 and one to D13, their other end connect to the ground. The switches select the scan rate and the trigger level.

```
//*****************************************
// Three color 5msps ext AtoD Scope
// By Bob Davis
// UTFT_(C)2012 Henning Karlsen
// web: http://www.henningkarlsen.com/electronics
//
// Switches on D12 & D13 determine sweep speed and trigger level
//*****************************************

#include <UTFT.h>
// Declare which fonts we will be using
extern uint8_t SmallFont[];
extern uint8_t BigFont[];
extern uint8_t SevenSegNumFont[];

// Note that the control pins are now assigned to 8-11
UTFT myGLCD(ILI9325C,8,9,10,11);
int Input=0;
byte Sample[320];
byte OldSample[320];
int StartSample=0;
int EndSample=0;
int MaxSample=0;
int MinSample=0;
int mode=0;
int dTime=1;
int Trigger=10;
int SampleSize=0;
int SampleTime=0;

void DrawMarkers(){
  myGLCD.setColor(0, 220, 0);
  myGLCD.drawLine(0, 0, 0, 240);
  myGLCD.drawLine(60, 0, 60, 240);
  myGLCD.drawLine(120, 0, 120, 240);
  myGLCD.drawLine(180, 0, 180, 240);
  myGLCD.drawLine(239, 0, 239, 240);
  myGLCD.drawLine(319, 0, 319, 240);
  myGLCD.drawLine(0, 0, 319, 0);
  myGLCD.drawLine(0, 60, 319, 60);
```

```
  myGLCD.drawLine(0, 120, 319, 120);
  myGLCD.drawLine(0, 180, 319, 180);
  myGLCD.drawLine(0, 239, 319, 239);
}

void setup() {
  myGLCD.InitLCD();
  myGLCD.clrScr();
  pinMode(12, INPUT);
  digitalWrite(12, HIGH);
  pinMode(13, INPUT);
  digitalWrite(13, HIGH);
  pinMode(14, INPUT);
  pinMode(15, INPUT);
  pinMode(16, INPUT);
  pinMode(17, INPUT);
  pinMode(18, INPUT);
  pinMode(19, INPUT);
}

void loop() {
// Set the background color(Red, Green, Blue)
  myGLCD.setBackColor(0, 0, 0);
  myGLCD.setFont(BigFont);
  char buf[12];
  while(1) {
    DrawMarkers();
    if (digitalRead(13) == 0) mode++;
    if (mode > 10) mode=0;
// Select delay times for scan modes
    if (mode == 0) dTime=0;
    if (mode == 1) dTime=0;
    if (mode == 2) dTime=1;
    if (mode == 3) dTime=2;
    if (mode == 4) dTime=5;
    if (mode == 5) dTime=10;
    if (mode == 6) dTime=20;
    if (mode == 7) dTime=50;
    if (mode == 8) dTime=100;
    if (mode == 9) dTime=200;
    if (mode == 10) dTime=500;
```

```
// Select trigger level
   if (digitalRead(12) == 0) Trigger=Trigger+10;
   if (Trigger > 50) Trigger=0;
// Wait for input to be greater than trigger
   while (Input < Trigger){
   Input = PINC;
   }

// Collect the analog data into an array
   if (mode == 0) {
// Read analog port as a parallel port no loop
   StartSample = micros();
   Sample[0]=PINC;
   Sample[1]=PINC;      Sample[2]=PINC;      Sample[3]=PINC;
   Sample[4]=PINC;      Sample[5]=PINC;      Sample[6]=PINC;
   Sample[7]=PINC;      Sample[8]=PINC;      Sample[9]=PINC;
   Sample[10]=PINC;     Sample[11]=PINC;     Sample[12]=PINC;
   Sample[13]=PINC;     Sample[14]=PINC;     Sample[15]=PINC;
   Sample[16]=PINC;     Sample[17]=PINC;     Sample[18]=PINC;
   Sample[19]=PINC;     Sample[20]=PINC;     Sample[21]=PINC;
   Sample[22]=PINC;     Sample[23]=PINC;     Sample[24]=PINC;
   Sample[25]=PINC;     Sample[26]=PINC;     Sample[27]=PINC;
   Sample[28]=PINC;     Sample[29]=PINC;     Sample[30]=PINC;
   Sample[31]=PINC;     Sample[32]=PINC;     Sample[33]=PINC;
   Sample[34]=PINC;     Sample[35]=PINC;     Sample[36]=PINC;
   Sample[37]=PINC;     Sample[38]=PINC;     Sample[39]=PINC;
   Sample[40]=PINC;     Sample[41]=PINC;     Sample[42]=PINC;
   Sample[43]=PINC;     Sample[44]=PINC;     Sample[45]=PINC;
   Sample[46]=PINC;     Sample[47]=PINC;     Sample[48]=PINC;
   Sample[49]=PINC;     Sample[50]=PINC;     Sample[51]=PINC;
   Sample[52]=PINC;     Sample[53]=PINC;     Sample[54]=PINC;
   Sample[55]=PINC;     Sample[56]=PINC;     Sample[57]=PINC;
   Sample[58]=PINC;     Sample[59]=PINC;     Sample[60]=PINC;
   Sample[61]=PINC;     Sample[62]=PINC;     Sample[63]=PINC;
   Sample[64]=PINC;     Sample[65]=PINC;     Sample[66]=PINC;
   Sample[67]=PINC;     Sample[68]=PINC;     Sample[69]=PINC;
   Sample[70]=PINC;     Sample[71]=PINC;     Sample[72]=PINC;
   Sample[73]=PINC;     Sample[74]=PINC;     Sample[75]=PINC;
   Sample[76]=PINC;     Sample[77]=PINC;     Sample[78]=PINC;
   Sample[79]=PINC;     Sample[80]=PINC;     Sample[81]=PINC;
   Sample[82]=PINC;     Sample[83]=PINC;     Sample[84]=PINC;
```

```
Sample[85]=PINC;    Sample[86]=PINC;    Sample[87]=PINC;
Sample[88]=PINC;    Sample[89]=PINC;    Sample[90]=PINC;
Sample[91]=PINC;    Sample[92]=PINC;    Sample[93]=PINC;
Sample[94]=PINC;    Sample[95]=PINC;    Sample[96]=PINC;
Sample[97]=PINC;    Sample[98]=PINC;    Sample[99]=PINC;
Sample[100]=PINC;   Sample[101]=PINC;   Sample[102]=PINC;
Sample[103]=PINC;   Sample[104]=PINC;   Sample[105]=PINC;
Sample[106]=PINC;   Sample[107]=PINC;   Sample[108]=PINC;
Sample[109]=PINC;   Sample[110]=PINC;   Sample[111]=PINC;
Sample[112]=PINC;   Sample[113]=PINC;   Sample[114]=PINC;
Sample[115]=PINC;   Sample[116]=PINC;   Sample[117]=PINC;
Sample[118]=PINC;   Sample[119]=PINC;   Sample[120]=PINC;
Sample[121]=PINC;   Sample[122]=PINC;   Sample[123]=PINC;
Sample[124]=PINC;   Sample[125]=PINC;   Sample[126]=PINC;
Sample[127]=PINC;   Sample[128]=PINC;   Sample[129]=PINC;
Sample[130]=PINC;   Sample[131]=PINC;   Sample[132]=PINC;
Sample[133]=PINC;   Sample[134]=PINC;   Sample[135]=PINC;
Sample[136]=PINC;   Sample[137]=PINC;   Sample[138]=PINC;
Sample[139]=PINC;   Sample[140]=PINC;   Sample[141]=PINC;
Sample[142]=PINC;   Sample[143]=PINC;   Sample[144]=PINC;
Sample[145]=PINC;   Sample[146]=PINC;   Sample[147]=PINC;
Sample[148]=PINC;   Sample[149]=PINC;   Sample[150]=PINC;
Sample[151]=PINC;   Sample[152]=PINC;   Sample[153]=PINC;
Sample[154]=PINC;   Sample[155]=PINC;   Sample[156]=PINC;
Sample[157]=PINC;   Sample[158]=PINC;   Sample[159]=PINC;
Sample[160]=PINC;   Sample[161]=PINC;   Sample[162]=PINC;
Sample[163]=PINC;   Sample[164]=PINC;   Sample[165]=PINC;
Sample[166]=PINC;   Sample[167]=PINC;   Sample[168]=PINC;
Sample[169]=PINC;   Sample[170]=PINC;   Sample[171]=PINC;
Sample[172]=PINC;   Sample[173]=PINC;   Sample[174]=PINC;
Sample[175]=PINC;   Sample[176]=PINC;   Sample[177]=PINC;
Sample[178]=PINC;   Sample[179]=PINC;   Sample[180]=PINC;
Sample[181]=PINC;   Sample[182]=PINC;   Sample[183]=PINC;
Sample[184]=PINC;   Sample[185]=PINC;   Sample[186]=PINC;
Sample[187]=PINC;   Sample[188]=PINC;   Sample[189]=PINC;
Sample[190]=PINC;   Sample[191]=PINC;   Sample[192]=PINC;
Sample[193]=PINC;   Sample[194]=PINC;   Sample[195]=PINC;
Sample[196]=PINC;   Sample[197]=PINC;   Sample[198]=PINC;
Sample[199]=PINC;   Sample[200]=PINC;   Sample[201]=PINC;
Sample[202]=PINC;   Sample[203]=PINC;   Sample[204]=PINC;
Sample[205]=PINC;   Sample[206]=PINC;   Sample[207]=PINC;
```

```
    Sample[208]=PINC;   Sample[209]=PINC;   Sample[210]=PINC;
    Sample[211]=PINC;   Sample[212]=PINC;   Sample[213]=PINC;
    Sample[214]=PINC;   Sample[215]=PINC;   Sample[216]=PINC;
    Sample[217]=PINC;   Sample[218]=PINC;   Sample[219]=PINC;
    Sample[220]=PINC;   Sample[221]=PINC;   Sample[222]=PINC;
    Sample[223]=PINC;   Sample[224]=PINC;   Sample[225]=PINC;
    Sample[226]=PINC;   Sample[227]=PINC;   Sample[228]=PINC;
    Sample[229]=PINC;   Sample[230]=PINC;   Sample[231]=PINC;
    Sample[232]=PINC;   Sample[233]=PINC;   Sample[234]=PINC;
    Sample[235]=PINC;   Sample[236]=PINC;   Sample[237]=PINC;
    Sample[238]=PINC;   Sample[239]=PINC;   Sample[240]=PINC;
  EndSample = micros();
  }
  if (mode == 1) {
// Read analog port as a parallel port no delay
  StartSample = micros();
  for(int xpos=0; xpos<240; xpos++) {
    Sample[xpos]=PINC;
  }
  EndSample = micros();
  }
  if (mode >= 2) {
// Read analog port as a parallel port variable delay
  StartSample = micros();
  for(int xpos=0; xpos<240; xpos++) {
    Sample[xpos]=PINC;
    delayMicroseconds(dTime);
  }
  EndSample = micros();
  }

// Display the collected analog data from array
  for(int xpos=0; xpos<239; xpos++) {
// Erase the old stuff
    myGLCD.setColor(0, 0, 0);
    myGLCD.drawLine (xpos+1, 255-OldSample[xpos+1]*4, xpos+2, 255-OldSample[xpos+2]*4);
    if (xpos==0) myGLCD.drawLine (xpos+1, 1, xpos+1, 239);
// Draw the new data
    myGLCD.setColor(255, 255, 255);
```

```
    myGLCD.drawLine (xpos, 255-Sample[xpos]*4, xpos+1, 255-Sample[xpos+1]*4);
  }

//  Determine sample voltage peak to peak
  MaxSample = Sample[100];
  MinSample = Sample[100];
  for(int xpos=0; xpos<240; xpos++) {
   OldSample[xpos] = Sample[xpos];
   if (Sample[xpos] > MaxSample) MaxSample=Sample[xpos];
   if (Sample[xpos] < MinSample) MinSample=Sample[xpos];
   }
// display the sample time, delay time and trigger level
  myGLCD.setColor(0, 0, 255);
  SampleTime=EndSample-StartSample;
  myGLCD.print("uSec.", 240, 10);
  myGLCD.print("    ", 240, 30);
  myGLCD.print(itoa(SampleTime, buf, 10), 240, 30);
  myGLCD.print("Delay", 240, 70);
  myGLCD.print("    ", 240, 90);
  myGLCD.print(itoa(dTime, buf, 10), 240, 90);
  myGLCD.print("Trig.", 240, 130);
  myGLCD.print(itoa(Trigger, buf, 10), 240, 150);
// Range of 0 to 64 * 78 = 4992 mV
  SampleSize=(MaxSample-MinSample)*78;
  myGLCD.print("mVolt", 240, 190);
  myGLCD.print(itoa(SampleSize, buf, 10), 240, 210);
    }
}
```

Bibliography

Programming Arduino
Getting Started With Sketches
By Simon Mark
Copyright 2012 by the McGraw-Hill Companies

This book gives a thorough explanation of the programming code for the Arduino. However the projects in the book are very basic. It does cover LCD's and Ethernet adapters.

Getting Started with Arduino
By Massimo Banzi
Copyright 2011 Massimo Banzi

This author is a co-founder of the Arduino.
This book has a quick reference to the programming code and some simple projects.

Arduino Cookbook
by Michael Margolis
Copyright © 2011 Michael Margolis and Nicholas Weldin. All rights reserved.
Printed in the United States of America.
Published by O'Reilly Media, Inc., 1005 Gravenstein Highway North, Sebastopol, CA.

This book has lots of great projects, with a very good explanation for every project.

Practical Arduino: Cool Projects for Open Source Hardware
Copyright © 2009 by Jonathan Oxer and Hugh Blemings
ISBN-13 (pbk): 978-1-4302-2477-8
ISBN-13 (electronic): 978-1-4302-2478-5
Printed and bound in the United States of America

Page 197 of this book has how to supercharge the Analog to Digital converter for faster sampling rates in oscilloscopes and logic analyzers.

Printed in Great Britain
by Amazon